Role of Artificial Intelligence in Higher Education

Aftab Ara
Editor

Role of Artificial Intelligence in Higher Education

Pathways to Sustainable Development

Springer

Editor
Aftab Ara
College of Business
University of Ha'il
Hail, Saudi Arabia

ISBN 978-981-95-5349-5 ISBN 978-981-95-5350-1 (eBook)
https://doi.org/10.1007/978-981-95-5350-1

This Springer imprint is published by the registered company Springer Nature Singapore Pte Ltd.
The registered company address is: 152 Beach Road, #21-01/04 Gateway East, Singapore 189721, Singapore

Preface

Artificial Intelligence (AI) has emerged as a new horizon of possibilities for higher education. With the world trying to achieve United Nation Sustainability Development Goals (SDG), particularly SDG 4: Quality Education, with the application of AI into higher education both as an opportunity and a challenge. These critical aspects are explored in the book, *Role of Artificial Intelligence in Higher Education: Pathways to Sustainable Development*.

The main objective of this book is to bring out AI's evolving role for a sustainable and inclusive education. The insights are diverse perspectives from academia, industry experts, and policymakers reflecting on how AI transform teaching, learning, research, and governance in higher education institutions. This book addresses the AI adoption through ethical considerations, accessibility, equity, and policy framework highlighting successful practices and ongoing challenges.

This edited book is not only a scholarly contribution but a guide for educators, researchers, decision-makers, and technology developers. It helps to improve critical dialogue for AI innovations with the principles of sustainability, by ensuring that technology is the main catalyst for equitable learning and global development.

We hope that this book will inspire readers to recognize higher education as an ecosystem which is future-proofed. AI needs to facilitate creativity, inclusivity, and resilience in higher education. Through the revolution of AI for sustainable development, this book hopes to contribute a useful perspective to global discussion regarding education and innovation in the 21st century.

We would like to add our gratitude to all of the authors, reviewers, and collaborators who have contributed their experience and expertise to this book. Their commitment and hard work have been instrumental in making this book a valuable resource for the global academic and working fraternity.

Dr. Aftab Ara
College of Business Administration
University of Hail
Hail, Saudi Arabia

Acknowledgements

While writing a book is a solitary journey, but it is made possible by the steadfast support of numerous people. I am thankful for the belief and generosity of those who have stood by me in writing my book—*Role of Artificial Intelligence in Higher Education: Pathways to Sustainable Development*.

To my family members I owe them for their endless love, patience, and encouragement that have been the foundation upon which this book was built. Thank you for giving me the space and grace to pursue this dream, and for never letting me forget the joy of the journey.

Thanks to all the authors of the book which have contributed their insights for this book.

Thank you to the team at Springer Nature for believing in this book and for bringing it to life.

Finally, to the readers—thank you. We wrote this for you, and are humbled and thrilled that you have chosen to spend time with these words. I hope this book brings you as much joy in reading it as it has brought me in writing it.

With deep appreciation.

Dr. Aftab Ara

Contents

AI-Enhanced Learning for Sustainable Development: Balancing Innovation and Equity in Higher Education

Aftab Ara

Abstract The SDGs, developed by the United Nations to address the global predicament, lay down SDG 4 on quality education. In Higher education, these SDGs are achieved by becoming knowledge-makers, knowledge-disseminators, and knowledge-implementers. However, discrepancies remain on various fronts, such as access, quality, and infrastructure across different geographical contexts. AI could create a disruptive impact in the educational space through personalized learning, improved efficiency, and data-driven strategies. The intersection between AI and SDG 4 can create opportunities to tackle the barriers of education and expedite the journey toward inclusive and equitable quality education. The chapter discusses how AI in education began historically, what the present climate looks like concerning AI integration into higher education around the globe, and the ensuing ethical considerations. The implementations examined revealed the successful use of AI for advancing the objectives of educational sustainability. The Digital divide, competition between AI optimization and human values, and privacy concerns are challenges to be recognized and examined. The requirements for responsible AI integration will include policy emergence, the forging of partnerships between public and private sectors, and the development of AI literacy among educators and students. Striking a balance between technological advances and cultural responsiveness and contextual pertinence will be fundamental to harnessing AI for supporting inclusive, effective, and sustainable learning environments towards the realization of SDG 4 and the wider SDG agenda.

Keywords Development goals (SDGs) · SDG 4: Quality education · Higher education · Artificial intelligence (AI) · Educational access · Educational equity · Educational quality · Sustainable development

---

A. Ara (✉)
College of Business Administration, University of Hail, Hail, Saudi Arabia

A. Ara (ed.), *Role of Artificial Intelligence in Higher Education*,
https://doi.org/10.1007/978-981-95-5350-1_1

1 Introduction

Artificial Intelligence (AI) is becoming one of the most significant developments in higher education today: it promises better access, equity, and quality in learning. Because much happens not only at a distance in geography but also in quality and infrastructure, the basic challenge to the whole potential that AI holds for education poses critical questions on: 'how can AI be put to effective use in overcoming these inequalities and creating conditions for all-inclusive educational environments'.

Internal inequalities in provision concerning access to education, as well as quality of education, form significant barriers that could be eliminated toward achieving the goal of Sustainable Development Goal 4 (SDG 4): to promote inclusive and equitable quality education. The rapid growth of AI technology should be done ethically.

The study aims to analyze the evolution of artificial intelligence in education, review the current status of AI in higher education globally, and investigate all ethical implications. Hence, the research aims to come up with the optimum strategies for the best utilization of AI in promoting equity in education.

This research aims to explore ways in which accountable AI integration can advance the potential of SDG 4 through examining the policy implications that would follow public-private partnerships, as well as the development of AI literacy among educators and students. There will be a focus on inclusive, efficient, and sustainable learning environments for attaining educational equality.

Study Implications.

This study can provide recommendations for policymakers, educational institutions, and other stakeholders concerning AI integration in education. The knowledge provided, informed by the ethical and procedural aspects of AI deployment, will help shape robust frameworks and guide requirements for responsible and equitable use of AI technologies. Ultimately, then, this study accounts for more equitable education through its explorations of such promises made possible through AI but increasingly complicated by ethical questions.

2 Literature Review

2.1 *The SDG Framework and Education's Central Role*

2.1.1 The Interconnectedness of the 17 SDGs and their Integrated Approach to Sustainable Development

The 17 SDGs of the UN 2030 Agenda as an indivisible framework, created by which the world is going to be made sustainable. From this perspective, it recognizes various intricate issues like poverty, hunger, health, education, and environmental degradation, which can only be targeted jointly because their nature is interrelated (Weiland et al., 2021). Thus, this integrated approach to sustainable development requires deliberate cooperation between all political levels and across various sectors by states and non-state actors. It has, however, been established that the links between goals are rather uneven, and full integration of gender equality, peace, and governance concerns has not been achieved (Mcgowan et al., 2018). Such incoherence may undermine policy initiatives intended for the implementation of the SDGs. There are, therefore, potential tradeoffs in trying to realize these goals at the same time. A systems approach to sustainability illustrates the three systems, that is, "the environmental system," "the economic system," and "the social system," in which development is sustainable. This systems approach allows reality-checking about welfare effects when progress in one SDG occurs to the detriment of another (Barbier & Burgess, 2017). In conclusion, the SDGs are highly connected as they are presented; however, a fruitful undertaking requires careful handling of overlap complexities. The COVID-19 pandemic further highlighted the interdependencies between SDGs and revealed their unique patterns of interconnectedness (Shulla et al., 2021). The nexus approach to corporate sustainability is one such approach that has been put forth to address these issues (Van Zanten and Van Tulder [2021]) and those within the Just Digital Ethical Framework (Clark et al., 2022). This will eventually be the litmus test for the success of the 2030 agenda(Verboven & Vanherck, 2016), which calls for new governance frameworks and integrated assessment approaches (Baumgartner, 2019; Weiland et al., 2021).

2.1.2 Positioning of SDG 4 as both a Standalone Goal and an Enabler for Achieving Other SDGs

Education is a basis for the accomplishment of other SDGs, and SDG4 in itself is thus not only a goal just for the sake of it but also to be rightly called an enabler in fulfilling other SDGs. The Agenda 2030 recognizes education as a key driver for sustainable development, with Target 4.7 specifically outlining Education for Sustainability as its area of focus (Giangrande et al., 2019). In this respect, there is an increased emphasis on the interlinkage that education has in fulfilling aspirations expressed in Agenda 2030 and the SDGs. Analysis by the semantic network indicates that all targets of the SDGs are related in various respects to operational goals, thus making the current organization of 17 SDGs to define and develop policies unacceptable. Intersectoral strategies ought to be implemented in communities in consideration of interlinkages among targets (Song & Jang, 2023). This underpins the pertinence of SDG 4 as an enabler, for education cuts across developmental goals. Thus, SDG 4 will bear a duality- the character of a goal in itself and proceed to other goals distinctive of its catalytic role. In this direction, we could cite the Times Higher Education Impact Ranking that ranks universities concerning the SDGs, showing how higher education contributed to the performance of many SDGs (Ashida, 2022). Alongside this, lifelong literacy is contained within the Education 2030 Framework for Action, literacy as a lifelong learning process, life-wide learning unto sector-wide reforms in the direction of lifetime learning systems (Hanemann, 2019). Therefore, this multifaceted view on education strengthens the position of SDG 4 as a goal in its own right and as a catalyst for attaining all other Sustainable Development Goals.

2.1.3 Evolution of Education Targets from the Millennium Development Goals to the more Comprehensive SDG Framework

From the MDGs to the SDGs, the transformation in the education targets involved going from one strategy or approach to a different one at a wider breadth. In its education target settings, the MDGs framed targets for education narrowly; they were mainly school-based and confined two targets to two of the goals (Unterhalter, 2014). Such limited-framed conditions had perverse ramifications, omitting crucial aspects of quality, context, and equity in education (Unterhalter, 2014). The SDGs, as contrasted with previous decades, embraced a much wider and integrated view of education. Hopefully, these SDGs and the framework of their indicators would become the basis for a more robust policy framework for global development (Janoušková et al., 2018). It's noteworthy that the outcomes from assessments of SDGs could be ambiguous and confusing unless supported by a well-designed conceptual indicator framework (Janoušková et al., 2018). As a whole, the SDGs are a much more integrated system than the MDGs, which can probably encourage cross-sector policy integration (Le Blanc, 2015). To sum up, the change from MDGs to SDGs in the education sector illustrates a broader transition toward a more holistic and interconnected approach to global development. In contrast to its narrow focus on school input-targets established by MDGs, SDGs defined education more broadly by addressing its interrelation with other development areas. Successful implementation and assessment mechanisms, as well as the potential to innovate ways to bridge different development goals, are vital in determining the success of the new framework (Le Blanc, 2015; Janoušková et al., 2018).

2.1.4 The Specific Targets within SDG 4 that Relate to Higher Education Access, Quality, and Outcomes

Related to higher education access, quality, and outcomes, particularly concerning active SDG 4, there are specific targets: Target 4.3 seeks to 'ensure equal access for all women and men to affordable and quality technical, vocational, and tertiary education, including university' by 2030 (Ashida, 2022). It initiates discussions on access and quality, underscoring that any higher education opportunities should be open equally to all genders (Ashida, 2022; Heleta & Bagus, 2020). While SDG 4 is a promoter of equal access to higher education, there are contradictions in its path, especially in the case of low-income countries. As Heleta and Bagus (2020) argue, "In its recommendations, the SDGs should help these countries strengthen their fragile higher education systems instead of, ironically, creating opportunities for select individuals to study at developed countries' universities. This approach would only perpetuate the dependence on foreign aid, coupled with a chronic capacity gap in these low-income countries." In sum, while SDG 4 presents lofty aspirations for higher education access and quality, the extent of implementation and impact varies from one context to another. The Times Higher Education Impact Ranking has pictured the global performance of universities about SDGs, with Japanese universities appearing active in addressing global issues (Ashida, 2022). However, beyond this, a more holistic approach is needed that views the challenges confronting low-income countries in developing higher education systems in a special light, to drive the attainment of inclusive and equitable quality education for all (Heleta & Bagus, 2020). (Ashida, 2022).

2.2 Higher Education as a Catalyst for Sustainable Development

2.2.1 The Unique Position of Higher Education Institutions as Knowledge Creators, Disseminators, and Implementers

Higher education institutions (HEIs) uniquely occupy themselves as knowledge creators, disseminators, and implementers because of their involvement in the transformation of societies to be more sustainable. HEIs, by these core functions, education, research, and outreach, essentially lead the trail towards sustainable development through further improvement and dissemination of knowledge creation and knowledge building; capacity building via training; and organization for increased resilience with hosts of local communities (Berchin et al., 2021). Interestingly, although they are major players in public science communication, HEIs are very few regarding studies on HEI communication practitioners in the universities and their public relations and outreach in communication. The four identified communicators in HEIs have diversified conceptions and professional characteristics in terms of role making, which illustrates diversity in approaches for knowledge dissemination across the institutions (Volk et al., 2023). Further, most importantly, the adoption of open educational resources (OER) in HEIs indicates great cost reductions and better learning experiences for students, but has been slow in coming to large-scale implementation (Annand & Jensen, 2017). Thus, HEIs are called upon to play increasingly active roles in the economic, social, and cultural aspects of development in their regions, besides their traditional roles of educating and researching (McCowan, 2023; Abowardah et al., 2024) (Singh & Blessinger, 2023) (Price et al., 2021) (Filho et al., 2024) (Trott et al., 2018) (Purcell et al., 2019) (Arbo & Benneworth, 2007). This unique position allows them to be both objects and subjects of societal change, adapting to external contexts while also creating positive impacts on society (Sporn & Godonoga, 2024). Such would mean that in the act of being knowledge creators, disseminators, and implementers, HEIs will be able to contribute positively toward sustainable development as well as regional development.

2.2.2 The Triple Mission of Universities: Education, Research, and Community Engagement in the Context of SDGs

Through education, research, innovation, and public engagement, universities are key to aiding in the realization of the United Nations Sustainable Development Goals (SDGs). The university's triple mission of teaching, research, and community engagement is pivotal to advancing the SDGs (McCowan, 2023). Among universities' primary functions, therefore, are educating ministry officials into sustainably engaged leaders and enabling them to raise awareness amongst and actively involve students to support the implementation of the SDGs (Abowardah et al., 2024). Therefore, aspects of ESD should be the core content of university curricula for developing sustainability competencies in students who are placed in leadership positions in a challenging world (Price et al., 2021). Research is of utmost importance in the university sector to address issues relating to sustainability. The university can use the "living labs" strategy for experimentation towards positioning sustainability intentionally, wherein the framework of SDG serves as an instrument to achieve that (Purcell et al., 2019). Collaborative research projects with or across universities could fast-track the delivery of SDGs (Purcell et al., 2019). Community engagement represents another key area of contribution by universities towards the SDGs. Therefore, the transformation of societies comes about through universities in collaboration with the local community, businesses, and other stakeholders (Purcell et al., 2019). Based on Participatory Action Research (PAR), Undergraduate Research Experiences (UREs) provide an action-oriented platform for universities to advance the SDG agenda while enhancing student growth and community cooperation (Trott et al., 2018). While full participation of universities in SDG implementation seems a good scenario, various obstacles stand in the way; these include challenges arising from a lack of financial resources, the burden of marketization, the rivalry for status, and difficulties related to assessment (McCowan, 2023). Some universities lack clarity on how they might prioritize the SDGs that can have the biggest impact, reduce their environmental footprint, and build meaningful collaborations (Singh & Blessinger, 2023). It can be concluded that the universities have more than enough potential through their triple missions towards the SDGs, but there is room for further improvement. Many higher education institutions have limited integration of ESD and engagement with the SDGs (Price et al., 2021). To increase the effectiveness of their contributions, universities need greater urgency in integrating ESD with teaching and learning, in overcoming institutional barriers, and in capitalizing on their unique position to steer transformational change of individuals and societies toward sustainable futures (Filho et al., 2024; Price et al., 2021).

2.2.3 Global Disparities in Higher Education Access, Quality, and Infrastructure

Differences in higher education access, quality, and infrastructure systems are significant barriers to global efforts toward achieving equitable education in the world.

Such disparities have become worse, especially concerning online learning resources, due to the COVID-19 pandemic (Hadjeris, 2021; Kara, 2021). While Goal 4 of the Sustainable Development Goals talks about ensuring inclusive and quality education for all by 2030, it seems apparent, judging by the responses of international education institutions during the pandemic, that meeting this challenge will be too difficult (Hadjeris, 2021). It is interesting to note that the expansion of higher education systems, like that of some countries such as China, did not guarantee a more equal distribution of educational opportunities among people. Equality of access continues to be an inequality factor conditioned by household registration, family economic status, and level of parental education (Wu et al., 2020). In addition, more interest in management, organization, and quality is shown in European higher education research, which may be at the expense of student access, formerly the most salient equity concern (Ramirez & Tiplic, 2013). In summary, remedying the global disparities in higher education will require a multi-dimensional approach, which will include the availability and accessibility of quality education as well as horizontal measures to protect disadvantaged students from studying in substandard institutions (McCowan, 2015). Such factors grow in complexity, evidencing the interplay of race, socioeconomic status, and resource access in equity improvement efforts (Allen, 2005; Darling-Hammond, 2004). In responding to these emerging challenges, an urgent need arises to emphasize blended learning, capacity building for online pedagogy, and holistic quality assurance frameworks for balancing learning outcomes with students' basic needs in higher education institutions (Kara, 2021).

2.2.4 Challenges Facing Higher Education in Developing Regions Versus Developed Economies

Higher education institutions in developing regions have unique challenges that distinguish them from those in developed economies. Most developing countries have weaknesses in institutional policies, deficiencies in infrastructure, overcrowded facilities, heavy teaching loads, and a lack of research resources, among other things (Abugre, 2017). Such fundamental hindrances delay academic development and the course of integrating sustainability into the curriculum and systems (Leal Filho et al., 2017). Interestingly, the challenge for developed countries is not about basic challenges like Internet connectivity and faculty training for online education, as exemplified by Pakistan during the COVID-19 pandemic, but affordability, access, and accountability (Farooq et al., 2020). Competition among institutions, which is generally observed in developed nations, is oftentimes not the best approach in developing nations (George, 2006). Thus, developing and developed regions differ in their challenges of higher education, although they share some. While developing countries need policies besides the basic establishment of infrastructure, most developed countries enhance existing systems for appropriate accountability and accessibility. It is essential that overcoming these challenges be done collaboratively with all stakeholders, institutionally supported, and through imagination (Farooq et al., 2020; Leal Filho et al., 2017). Tailoring governance models to the specific needs and realities of each (Abbas et al., 2023) (Maity & Deroy, 2024) country's higher education sector is also part of sustainable development (Abugre, 2017; George, 2006).

2.3 The AI Revolution in Education

2.3.1 Fundamental Concepts and Capabilities of AI Relevant to Educational Contexts

Different AI technologies in educational environments provide the fundamental conceptualizations and functions pertinent to learning-changing capabilities. Personalized learning systems, intelligent tutoring systems, and automated assessment systems have all become significant applications through which learning outcomes, engagement, and overall academic performance are enhanced (Abbas et al., 2023). These systems can interpret machine learning, natural language processing, and data mining algorithms into personalized instruction designed to address the unique needs of individual students (Rizvi, 2023). Generative AI, particularly large language models such as GPT-4, is changing the field of educational technology by allowing for content generation, feedback, and learning pathways in real time (Maity & Deroy, 2024). Automated question generation, personalized feedback systems, and interactive dialogue systems are included in such an approach; however, addressing the trade-offs resulting from the perceived pedagogical accuracy, biases in the models, and learner engagement remains a challenge (Bai, 2024; Maity & Deroy, 2024). The integration of AI in education brings along a wide range of advantages. Such advantages in teaching include individualized teaching, data-driven insight, and increased student engagement (Yılmaz, 2024). Although AI applications such as ChatGPT and intelligent tutoring systems do carry a great deal of promise toward creating better instruction and adaptive learning which results in improved educational outcomes(Zekaj, 2023) ultimately, the use of AI may help make more efficient use of human efforts at teaching, alongside real concerns of dehumanizing, data privacy, and disempowerment effects (Bai, 2024). AI, along with its abilities in an educational context, refers mostly to such tenets as personalization, adaptability, and effectiveness. While great expectations might be drawn toward positive changes in education through AI, the challenges of ethical considerations, data privacy, and the need for balanced policy interventions to leverage the reach of AI equitably and responsibly remain (Bai, 2024; Yılmaz, 2024; Zekaj, 2023).

2.3.2 Historical Development of AI Applications in Education, from Simple Automation to Adaptive Learning

The historical evolution of AI applications in education has dramatically changed from simple automation to intelligent adaptive learning systems. Thus, the initial status of AI in education was as assistive functionalities, which were especially targeted at automating simple tasks (Cheng & Liang, 2023). With the new trends in technology, the revolutionary early applications were then expanded by attaching multimedia and interactive features to their set of capabilities, preparing them for future advanced tools in Kai-enabled education (Cheng & Liang, 2023). The evolution of AI in education has thus been defined by a number of milestones. The earlier implementations were very basic in terms of administrative task automation and simple content delivery. Over time, as capabilities matured, applications developed into advanced areas such as personalized learning platforms, adaptive assessment tools, and virtual tutors (Aggarwal et al., 2023; Gaur et al., 2024). This process produced Intelligent Tutoring Systems (ITS) based on large language models such as GPT-4 for highly personalized and adaptive learning environments (Maity & Deroy, 2024). Most interestingly, even though the focus of artificial intelligence has been on cognitive development and vocational guidance, advances have extended its application within schools to embrace social and emotional learning as well (Cheng & Liang, 2023). This direction emphasizes how more people are now recognizing the tremendous possibilities of AI in addressing different educational needs. In addition, raising ethical issues has surfaced with the integration of AIs in education, which includes compromise to data privacy, algorithmic bias, and the risk of intensifying educational inequalities (Gaur et al., 2024).

2.3.3 Current State of AI Integration in Higher Education Globally

The incorporation of AI into higher education around the globe is rapidly on its way to mass action as institutions are beginning to recognize the capability of AI to transform the traditional conceptualization of teaching, learning, and administrative business processes. Some areas into which AI is being integrated are personalized learning, predicting students' performances, and administrative efficiency in higher education (Leffia et al., 2024; Magrill & Magrill, 2024). In colleges across the US, the predicted AI market growth between 2018–2022 is about 48%. Research has proven that the current use of AIEd leads to positive outcomes like improved learning outcomes, increased access and retention, lower education costs, and decreased time to completion (Hutson et al., 2022). AI is becoming a global necessity tool in education with personalization and localization of learning (López-Chila et al., 2023). Adoption processes may vary from country to country, with some countries, such as China and the United States, among the leaders in the field of AI development and its actual research and application, while others are still in the early stages of adoption (López-Chila et al., 2023). For example, in Nigeria, AI is on the fast-growing demand in the educational system today, but more mature techniques of AI, employed in other countries, are not yet extensively practiced (Bali, 2024). Likewise, Saudi Arabia is considering AI applications in learning, touching the surface of what such technologies can offer to educational problems and renewing the delivery of education (Alotaibi and Alshehri, 2023). The current state of AI integration in higher education globally is characterized by numerous opportune and challenging openings. AI promises value additions in terms of personalized learning as well as improved administrative efficiency and student experiences. However, it also has potential downside risks with regard to privacy, data security, ethics, and the need to train faculty in these technologies (Bobula, 2024; Leffia et al., 2024). The increased exploration into AI technology applications continues to spur the creation of guidelines involving assessment policy updates and stamped preparation for developing AI skills towards responsible and effective technology uptake (Bobula, 2024). In summary, the global integrative landscape of AI in higher education is indeed diverse and continuing to develop. Some countries and institutions head the list of AI adopters, while others work to catch up. Its transformative potential is widely recognized in education, yet successful implementation calls for attention to many problems and ethical considerations. Therefore, research and collaboration should be continued as the field advances to maximize such opportunities.

2.3.4 Ethical Considerations Surrounding AI Deployment in Educational Settings

The careful consideration of the ethical implications will be very important in opening up the potential and challenges that AI integration in educational settings can bring. AI technologies can dramatically change how education has traditionally been provided, delivering personalization of the learning experience and increasing opportunities for reducing administrative loads (Eden et al., 2024). The ethical issues concerning these innovations are quite significant, and they must be considered. One major ethical concern among many is the protection of data and privacy for students. This concern arises as educational institutions collect and analyze vast amounts of data about students for educational purposes in education, resulting in data-security and privacy-related issues (Eden et al., 2024). Data privacy and governance have been placed by the European Commission as an important aspect of AI deployment in the Digital Education Action Plan 2021–2027 (Morandín-Ahuerma, 2024). In this spirit, it was proposed to formulate a manifesto promoting seven key principles for the development of safe AI in education, including the protection of students' data and ethical transparency (García Peñalvo et al., 2024). The other ethical concern is possible algorithmic bias, which will intensify existing inequalities in society. People are generally concerned that the AI algorithms can reinforce some biased or unequal conditions (Eden et al., 2024). The other issue of concern is the digital divide, which can further worsen the already-existent inequity in education-related aspects (Ma & Jiang, 2023). Thus, guidelines to embed ethics during the implantation of AI, transparency in algorithms, and an inclusive design will go a long way in mitigating risks (Alali & Wardat, 2024). Another consideration is the ethical use of AI in assessment and decision-making cases. The face of ethical concern is about the student's performance assessment and making consequential decisions with AI (Eden et al., 2024). Stakeholders need to work together to see that responsible AI deployment in higher education maximizes its benefits (Slimi & Villarejo Carballido, 2023). Ongoing dialogue among educators, policymakers, technologists, and students is crucial to steer responsible AI integration in education (Alali & Wardat, 2024). AI can develop and revolutionize many areas of practice in education. However, it will be important to integrate education adequately and ethically with the future technology threat to maximize benefits from AI.

2.4 *Convergence of AI and SDG 4*

2.4.1 The Potential of AI

AI technologies can produce a high volume of very personalized learning experiences to enhance access to quality education. Such technologies can redirect resource use, making it targeted and ensuring that equal access to education that has not been sufficiently utilized is revalidated. AI can accomplish educational equity by removing barriers and providing lifelong learning opportunities to all. In the context of education, LMs do predictive analytics to identify students at risk, can be utilized to build personalized tutoring systems, and can offer several tools for academic automation, thereby allowing teachers to do what they can do best: teach (Artyukhov et al., 2024; Nedungadi et al., 2024). Interestingly, benefits aside about AI, it brings forth some challenges as identified by Motsuka et al. (2024), including ethical questions, data privacy, and increasing the digital divide. Careful management of AI integration will demand that AI serves as a tool promoting inclusive and sustainable education rather than deepening the already-large inequalities (Artyukhov et al., 2024). Moreover, impetus is needed to lay down robust frameworks that work via AI's advantages in promoting academic integrity, given the known challenges AI raises for academic misconduct like plagiarism (Artyukhov et al., 2024). It is safe to say that AI has its role to play in achieving specific targets in SDG 4, providing enough leverage to offer educational access, equity, and quality. However, this will require a co-construction ethos that binds technological innovation squarely to the SDG 4 objectives (Nedungadi et al., 2024). Future research should be directed at the long-term societal effects of AI integration in educational settings and the development of policy tools for enabling learning environments that welcome AI innovation but adhere to ethical standards and respect for societal interests (Alsagri & Sohail, 2024; Artyukhov et al., 2024).

2.4.2 AI for Educational Inclusion

In education, socioeconomic barriers to equitable opportunities, enhancing learning outcomes, and ensuring the inclusion of multiple backgrounds can be reworked with the help of AI(Edeni et al., 2024). The digital divide and earlier social inequalities impede the wide-scale adoption of AI in education. Failure to address these could only lead to greater educational inequity and inadequacy in the distribution of resources (Bulathwela et al., 2024). In India, for instance, AI in education could have created wonders, but the reasons infrastructure bottlenecks, resource constraints, and privacy concerns stood in the way(Agarwal & Vij, 2024). To surmount the barriers and maximize AI for educational inclusion, it requires a multifocal approach that passes through developing culturally responsive policies, removing algorithmic bias, and securing data privacy (Bulathwela et al., 2024; Davoodi, 2024). There is a way AI can strengthen inclusion and autonomous learning through tying right to the principles of Universal Design for Learning (UDL) through multiple means of representation, engagement, and expression (Saborío-Taylor & Rojas-Ramírez, 2024). This combination allows AI to provide egalitarian access to education using free participatory resources such as Wikipedia and Open Educational Resources (Bulathwela et al., 2024). Concentrating on these areas will ensure that AI becomes a robust agent for inclusion into education that transcends geographical, socioeconomic, and ability divides.

2.4.3 Case Studies of Successful AI Applications

AI applications are found in the AI literacy-funded project at Queen Mary University of London, to develop a teaching and learning toolkit to support educators in the application of AI into their curriculum (Zhou & Schofield, 2024).AI curriculum tools have been used to help students learn about future sustainability challenges and improve innovation in research (Khan et al., 2025). The use of ChatGPT in blended learning systems has also shown improved learning outcomes, enhanced student engagement, and personalized learning experiences (Alshahrani, 2023). In summary, these case studies depict AI to be an important catalyst toward achieving educational sustainability goals. With regard to administrative purposes such as content delivery and creating personalized learning experiences, AI technologies are helping build more inclusive, efficient, and sustainable educational systems (Alshahrani, 2023, and Khan et al., 2025). Work, while ethical consideration and challenges such as data protection and algorithmic bias also had to be countered, for the responsible and equitable implementation of AI into education.

2.5 Transformative Potential and Future Directions

2.5.1 Emerging Models for AI-Enhanced Higher Education

Although AI technology is transforming education systems all around the world, it is one of the most important enablers for Sustainable Development Goals (SDGs) in higher education. The new emerging models for AI-influenced high education are engines for sustainable development in many ways such as AI-initiated innovations in higher education, which can considerably have improving educational outcomes, resource management, and future sustainability (Khan et al. 2025): for instance, personalizing learning experiences, enhancing energy efficiency, and minimizing environmental impacts would lead to a much more sustainable and inclusive academy. This is aligned with the complete objective of revolutionizing education through AI, which makes scalable, individualized learning experiences possible and improves access to quality education (Alsagri & Sohail, 2024). Integration of AI into higher education will also prove to be challenging. Ethical issues such as data privacy, algorithmic bias, and digital inequality are important in the establishment of strong ethical frameworks and institutional policies (Khan et al., 2025). Such issues reflect the greater widespread challenges experienced in the execution of AI in sustainable development, like being open, safe, and ethical (Fan et al., 2023). Therefore, while AI has great prospects for making improvements in higher education regarding sustainability, it would require careful implementation. Future research will develop AI literacy programs (Yeter et al., 2024), while others will delve into ethical issues and consider hybrid AI models that will fit dynamic and uncertain environments in education (Chen et al., 2024). This way, higher education will convert into AI utilizers in benefiting toward a sustainable future, making the paradigm shift meaningful for students who are preparing for future challenges (Alsagri & Sohail, 2024; Yeter et al., 2024; Khan et al., 2025; Chen et al., 2024; Fan et al., 2023).

2.5.2 Policy Considerations for Responsible AI Integration in Educational Systems

Different facets are to be taken into account under a policy for the responsible integration of AI into educational systems corresponding to its ethicomoral, technical, and societal dimensions. The European Union's Digital Education Action Plan 2021–2027 outlines fundamental principles, along the lines of human agency and human oversight, transparency, inclusion, data privacy, technical robustness, and accountability (Morandín-Ahuerma, 2024). These provide a means of laying a framework from which comprehensive policy can be developed concerning the responsible application of AI in education. One factor firmly secured on the policy agenda would be within the ethical domain of AI in educational measurement. The AIME group aims to establish ethical standards to address concerns such as algorithmic bias, transparency, and fairness in the context of AI technologies for assessment (Bulut & Beiting-Parrish, 2024). Given the fact that such policies aim at the mitigation of these risks to ensure an equitable result for all students is, by itself, an interesting point, it must also be emphasized that such integration of AI into education raises concerns of digital poverty and possibly the enhancement of preexisting inequalities (Blonder & Feldman-Maggor, 2024). Policies ought to enshrine inclusive access to AI-enhanced educational resources and help ameliorate the digital divide with such provisions. Furthermore, according to this UNESCO framework on GenAI on education, cultural diversity and human agency must be incorporated into the policy considerations (Blonder & Feldman-Maggor, 2024). Summarily, effective policies for responsible AI integration into educational systems need to give preference to the ethical guide, seek technical resolution, and deal with inclusivity. This multiparty cooperation must bring together educators, policymakers, technologists, and ethicists in the process of policy formulation (Barnes & Hutson, 2024). To ensure appropriate and equitable application of AI in education, to leverage its promise in improving learning outcomes and experiences, a robust framework for integrating ethical consideration throughout the lifecycle of AI systems must be established(Morandín-Ahuerma, 2024) (Barnes & Hutson, 2024) (Bulut & Beiting-Parrish, 2024) (Blonder & Feldman-Maggor, 2024).

2.5.3 The Role of Public-Private Partnerships in Scaling AI Solutions for Education

Public-private partnerships (PPPs) are essential for scaling AI solutions for education and providing innovative approaches to enhancing access to, quality of, and efficiency within educational systems. Using the strengths from both public and private sectors, these partnerships attempt to address challenges and use technology as a tool for the advancement of education (Barrera-Osorio et al., 2012). Concerning AI implementation, PPPs will assist in overcoming barriers such as limited resources, expertise, and infrastructure within public institutions. Educational organizations can work with private entities and gain access to groundbreaking AI technologies, data analysis capabilities, and specialized knowledge (Mikhaylov et al., 2018). Such collaboration goes a long way in helping resource-constrained institutions and developing countries, where AI in education faces immense challenges due to a lack of equipment, manpower, expertise, and infrastructure (Mollura et al., 2020). Important to mention is that the PPP in support of AI implementation in education has its risks and challenges, including issues surrounding privacy of data, existing regulatory barriers, and the call for transparency in AI algorithms (Murdoch, 2021; Ochuba et al., 2024. A wider road exists for PPPs to scale AI solutions in educational contexts, particularly in resource-constrained setups. To ensure that the maximum benefits from the partnership are derived, PPPs should build capacity, avail of rights systems favoring their own goals, and make sure AI implementations are aligned with educational goals and ethical standards promoted by the community (Malerbi et al., 2023; Mikhaylov et al., 2018).

2.6 *Critical Perspectives and Challenges*

2.6.1 Digital Divides and Technological Dependencies that May Exacerbate Educational Inequalities

Digital technology can aggravate existing educational inequalities or lessen them. They would provide an opportunity for more personalized learning and an increased ability to access resources. On the other hand, they could widen the digital divide and deepen existing inequalities (Capraro et al., 2024). Digital divide is not discrimination based on one's given name; it brings in other forms of inequity to include education, income, gender, and one's generation status. These are some other factors that play important roles in holding back against digital equity, especially concerning middle-aged to older adults (Fang et al., 2018). Such multidimensional characteristics can serve in the education field, concerning the deepening socioeconomic cleavages and the historic colonization of the global North over the global South, creating systematic digital inequalities in the design and implementation of online learning (Han & Kumwenda, 2024). About 20% of college students who, as a group, have a better-than-average level of connectivity, nevertheless find it difficult to maintain their access to technology, other to such problems as having broken hardware, limits on their internet data, or connectivity problems. Students from lower socioeconomic backgrounds and students of color are disproportionately affected by these problems, and dependence on poorly functioning laptops is correlated with lower GPAs (Gonzales et al., 2018). This shows how much tech reliance matters in the educational sphere and how it can either enhance the inequalities or address them. All these approaches would best be combined with the design of online education focused on equity in conjunction with local input and specialized expertise to guarantee relevance and sustainability in various contexts (Han & Kumwenda, 2024). Furthermore, health and digital literacy programs can monitor access, use, and impact across society, to ensure that digital technologies mitigate rather than exacerbate existing educational inequalities (Azzopardi-Muscat & Sørensen, 2019).

2.6.2 AI Optimization and Humanistic Educational Values

The confluence of technology and humanistic values that is brought about by the introduction of AI into education occupies a complex space, with its resultant tensions needing careful engagement. AI has potential transformational possibilities for the area of learning: personalization, self-paced learning, and improved efficiency (Grubaugh et al., 2023; Qian, 2023). However, such optimization often conflicts with conventional humanistic educational values. AI may offer personalized pathways in education, but may have difficulty in dealing with the uncertainties that come with education, which are very important since in anatomy, human variation and practice nuances from healthcare professionals are important (Lazarus et al., 2022). In addition, the application of AI tools like ChatGPT in writing education introduces inquiries on learning and cheating, thus not an organized statement on the productivity of AI as a time-shortening agent in learning (Mah et al., 2024). Interestingly, the impact of these tensions on different groups is not uniform. Students in technical and humanities specializations have different views on the impact of AI in society (Gherheş & Obrad, 2018). On top of that, there is a different perception between urban and rural students regarding AI requirements, stressing the need for cultural competence and ethics in education about AI (Sanusi & Olaleye, 2022). The balanced approach is thus required to address the above-mentioned essences. These include making AI visible in integration by transparency, diversity in the developer, uncertain and variable use in AI, and extending beyond replacement to rather use AI to amplify the human (Lazarus et al., 2022). Further, incorporating cultural components, humanistic thinking, and ethical perspectives in designing AI content can serve to bring technology and education closer to each other (Grubaugh et al., 2023; Sanusi & Olaleye, 2022). Ultimately, efforts must be made to reap the fruits that AI can bring while retaining the basic humanistic elements of education to meet all stakeholders through a more open, effective, and adaptive learning paradigm.

2.6.3 Data Privacy, Sovereignty, and Security Considerations in Educational AI

AI in education raises concerns with regard to data privacy, data sovereignty, and data security. The Digital Education Action Plan 2021–2027 of the European Commission enforced the need for transparency, free from bias, and respect for privacy for data management. Ethical principles include human agency, inclusion, and accountability (Morandín-Ahuerma, 2024).AI in education poses concerns such as data breaches, algorithmic bias, and misuse of sensitive information (Ismail, 2024). Protecting student data against privacy violations needs immediate attention with AI in education (Huang, 2023). The highly data-reliant nature of AI systems, compounded by the very nature of privacy implications, raises serious concerns for the possible threats of surveillance and discrimination (Gilbert & Gilbert, 2024). Intriguingly enough, whereas privacy-imposing arrangements are often promoted as solutions, AIs may use them to acquire more data, thus furthering surveillance architecture under the guise of compliance with regulations (Yew et al., 2024). Thus, a multifaceted, comprehensive strategy must be laid out. Educational institutions must create unambiguous policies for the use of data, while policymakers must update regulations to capture all the complexities that come with AI. EdTech developers should follow a privacy-by-design rationale, and educators should undertake efforts to enhance their data literacy (Ismail, 2024). Additionally, new privacy-preserving AI technologies, such as federated learning and differential privacy, provide potential solutions for meeting privacy protection and innovation. Finally, in AI governance in education, we need a concerted set of actions that rely on technological, ethical, and regulatory means to ensure that AI technologies respect privacy, exercise data sovereignty, and meet security standards.

2.6.4 Balancing Technological Innovation with Cultural Responsiveness and Contextual Relevance

Technology-driven innovation is fundamental to promoting sustainable economic development and green transition (Li & Hu, 2021). However, sustainable and inclusive growth is guaranteed when technological innovation is balanced with cultural sensitivity and contextual relevance. Socio-tech ventures realize social value from innovation and struggle to maintain the balance between social and market strategic orientations (Scillitoe et al., 2018). The technology venture status increases the complexity of balancing these orientations and necessitates an enriched meaning of innovation with these cultural and contextual factors. Interestingly, cultural responsiveness transcends commerce. Special attention needs to focus on broad literacy problems that consider global economic, social, historical, and cultural issues in language learning and teaching due to increasing examples of multilingual, multiethnic, and multicultural classrooms in education (Porto, 2009). As such, cultural responsiveness is important across various industry sectors, particularly those involving technology. Balanced technological advancement and cultural sensitivity are, thus, crucial in technology innovation. The balance is attained through contextual ambidexterity that is entrenched in organizational culture and is critical for new product innovation and organizational development, particularly for high-tech firms in rapidly changing business settings (Wang & Rafiq, 2012). By integrating cultural and contextual factors into the equation, innovation is stimulated in ways that are culturally meaningful for businesses, especially in minority and vulnerable communities (Rio & Saligan, 2023). Thus, this is an approach that additionally makes the technological innovations effective and relevant for acceptance in multiculturally different settings.

3 Discussion

The entry of AI into higher education presents huge opportunities and challenges, especially with an eye to ensuring equity and quality in terms of educational access. The paper mentions that AI could influence educational outcomes by addressing the socioeconomic barriers and improving institutional resource allocation, especially in developing contexts. This is well-aligned with Sustainable Development Goals (SDGs), especially with SDG 4, which talks about inclusive and equitable quality education.

However, the paper also states that there should be a strong co-construction ethos linking the innovation of technology to the measurement of ethical standards and societal interest. This becomes necessary due to the chances of unintended effects of the rapidly adopted AI technologies, either intensifying existing inequalities or introducing bias in educational assessments. Concerns about the ethical consideration of AI are needed to handle the integrity of academia and the privacy of data to put AI as a tool that empowers and does not divide.

While the paper discusses the pros of AI in detail, it fails to provide empirical evidence for its claims. The argument and practice in AI would be enriched for policymakers and educational institutions if the paper included actual case studies describing successful AI in education. Furthermore, biases and ethical challenges in AI need more discussion for a balance in the impact on education.

The study guides educational institutions to promote AI technology in such a way that equity goals are achieved with all stakeholders' decisions to create an inclusive learning environment. Hence, using AI in education can improve equity and quality by checking ethical implications. More research should focus on empirical studies and policies for responsible AI in the education sector.

4 Conclusion

Artificial Intelligence (AI) in higher education is a huge opportunity to foster access, equity, and quality in education, with the Sustainable Development Goals (SDGs). However, AI can be used only with a careful balance between innovation and ethical considerations.

AI is one of the key mechanisms that can solve most large-scale problems in education, including resource allocation and violations of integrity. At the same time, AI has its downfalls, as it could perpetuate and exacerbate existing inequalities and introduce biases. This is why educational institutions should have strong frameworks and practical guidelines for facilitating responsible use of AI in the education space in a manner that also embraces equity.

Future research should focus on empirical investigations regarding the social effects of AI in any context of education in the longer term and the creation of toolkits for policy implementation and ethical AI practices. If these considerations are kept in focus, the stakeholders could realize AI as a propelling agent for establishing inclusive and equitable educational environments, and thereby for the attainment of SDGs.

References

Abbas, N., Hussain, M. H. A. I., Manzoor, R., Hussain, T., & Ali, I. (2023). Role of artificial intelligence tools in enhancing students' educational performance at higher levels. *Journal of Artificial Intelligence, Machine Learning and Neural Network, 35*, 36–49. https://doi.org/10.55529/jaimlnn.35.36.49

Abowardah, E. S., Nurunnabi, M., Aboelnagah, H., & Labib, W. (2024). Students perception of sustainable development in higher education in Saudi Arabia. *Sustainability, 16*(4), 1483. https://doi.org/10.3390/su16041483

Abugre, J. B. (2017). Institutional governance and management systems in sub-Saharan Africa higher education: Developments and challenges in a Ghanaian research university. *Higher Education, 75*(2), 323–339. https://doi.org/10.1007/s10734-017-0141-1

Agarwal, P., & Vij, A. (2024). Assessing the challenges and opportunities of artificial intelligence in Indian education. *International Journal for Global Academic & Scientific Research, 3*(1), 36–44. doi:https://doi.org/10.55938/ijgasr.v3i1.71

Aggarwal, D., Sharma, D., & Saxena, A. B. (2023). Adoption of artificial intelligence (AI) for development of smart education as the future of a sustainable education system. *Journal of Artificial Intelligence, Machine Learning and Neural Network, 36*, 23–28. https://doi.org/10.55529/jaimlnn.36.23.28

Alali, R., & Wardat, Y. (2024). Opportunities and challenges of integrating generative artificial intelligence in education. *International Journal of Religion, 5*(7), 784–793. https://doi.org/10.61707/8y29gv34

Allen, W. R. (2005). A forward glance in a Mirror: Diversity challenged—Access, equity, and success in higher education. *Educational Researcher, 34*(7), 18–23. https://doi.org/10.3102/0013189x034007018
Alotaibi, N. S., & Alshehri, A. H. (2023). Prospers and obstacles in using artificial intelligence in Saudi Arabia higher education institutions—The potential of AI-based learning outcomes. *Sustainability, 15*(13), 10723. https://doi.org/10.3390/su151310723
Alsagri, H. S., & Sohail, S. S. (2024). Evaluating the role of artificial intelligence in sustainable development goals with an emphasis on quality education. *Discover Sustainability, 5*(1). https://doi.org/10.1007/s43621-024-00682-9
Alshahrani, A. (2023). The impact of ChatGPT on blended learning: Current trends and future research directions. *International Journal of Data and Network Science, 7*(4), 2029–2040. https://doi.org/10.5267/j.ijdns.2023.6.010
Annand, D., & Jensen, T. (2017). Incentivizing the production and use of open educational resources in higher education institutions. *The International Review of Research in Open and Distributed Learning, 18*(4). https://doi.org/10.19173/irrodl.v18i4.3009
Arbo, P., & Benneworth, P. (2007). Understanding the regional contribution of higher education institutions. *Organisation for economic co operation development*. https://doi.org/10.1787/161208155312
Artyukhov, A., Vasylieva, T., Artyukhova, N., Wołowiec, T., & Bogacki, S. (2024). SDG 4, academic integrity and artificial intelligence: Clash or win-win cooperation? *Sustainability, 16*(19), 8483. https://doi.org/10.3390/su16198483
Ashida, A. (2022). *The role of higher education in achieving the sustainable development goals* (pp. 71–84). springer nature singapore. https://doi.org/10.1007/978-981-19-4859-6_5
Azzopardi-Muscat, N., & Sørensen, K. (2019). Towards an equitable digital public health era: Promoting equity through a health literacy perspective. *European Journal of Public Health, 29*(Suppl 3), 13–17. https://doi.org/10.1093/eurpub/ckz166
Bai, X. (2024). The role and challenges of artificial intelligence in information technology education. *Pacific International Journal, 7*(1), 86–92. https://doi.org/10.55014/pij.v7i1.524
Bali, B. (2024). *Analysis of Emerging Trends in Artificial Intelligence in Education in Nigeria*. Research square platform llc. doi:https://doi.org/10.21203/rs.3.rs-3819828/v1
Barbier, E. B., & Burgess, J. C. (2017). The sustainable development goals and the systems approach to sustainability. *Economics, 11*(1). https://doi.org/10.5018/economics-ejournal.ja.2017-28
Barnes, E., & Hutson, J. (2024). Navigating the ethical terrain of AI in higher education: Strategies for mitigating bias and promoting fairness. *Forum for Education Studies, 2*(2), 1229. https://doi.org/10.59400/fes.v2i2.1229
Barrera-Osorio, F., Guaqueta, J., & Patrinos, H. A. (2012). *The role and impact of public private partnerships in Education*. Edward elgar. https://doi.org/10.4337/9780857930699.00019
Baumgartner, R. J. (2019). Sustainable development goals and the Forest sector—A complex relationship. *Forests, 10*(2), 152. https://doi.org/10.3390/f10020152
Berchin, I. I., De Aguiar Dutra, A. R., & Guerra, J. (2021). How do higher education institutions promote sustainable development? *A literature review. Sustainable Development, 29*(6), 1204–1222. https://doi.org/10.1002/sd.2219
Blonder, R., & Feldman-Maggor, Y. (2024). AI for chemistry teaching: Responsible AI and ethical considerations. *Chemistry Teacher International, 6*(4), 385–395. https://doi.org/10.1515/cti-2024-0014
Bobula, M. (2024). Generative artificial intelligence (AI) in higher education: A comprehensive review of challenges, opportunities, and implications. *Journal of Learning Development in Higher Education, 30*. https://doi.org/10.47408/jldhe.vi30.1137
Bulathwela, S., Pérez-Ortiz, M., Holloway, C., Shawe-Taylor, J., & Cukurova, M. (2024). Artificial intelligence alone will not democratise education: On educational inequality. *Techno-Solutionism and Inclusive Tools. Sustainability, 16*(2), 781. https://doi.org/10.3390/su16020781

Bulut, O., & Beiting-Parrish, M. (2024). The rise of artificial intelligence in educational measurement: Opportunities and ethical challenges. *Chinese/English Journal of Educational Measurement and Evaluation, 5*(3). https://doi.org/10.59863/miql7785

Capraro, V., Van Bavel, J. J., Bilancini, E., Van Der Linden, S., Hashimoto, D. A., Jetten, J., Douglas, K. M., Suri, S., Lentsch, A., Sutcliffe, J., Brañas-Garza, P., Viale, R., Rahwan, I., Tomlinson, J., Wall, F., Longoni, C., Lunn, P., Singh, V., Van Bavel, J. J., & Butera, L. (2024). The impact of generative artificial intelligence on socioeconomic inequalities and policy making. *PNAS Nexus, 3*(6). https://doi.org/10.1093/pnasnexus/pgae191

Chen, W., Juan, A. A., Fuster, N., Men, Y., & Osorio, C. (2024). Artificial intelligence in logistics optimization with sustainable criteria: A review. *Sustainability, 16*(21), 9145. https://doi.org/10.3390/su16219145

Cheng, Y., & Liang, Y. S. (2023). The development of artificial intelligence in career initiation education and implications for China. *European Journal of Artificial Intelligence and Machine Learning, 2*(4), 4–10. https://doi.org/10.24018/ejai.2023.2.4.32

Clark, S., Boyle, N., Carroll, C., Maclachlan, M., Hand, K., Marshall, K., Morahan, N., & O'Sullivan, K. (2022). Including digital connection in the United Nations sustainable development goals: A systems thinking approach for achieving the SDGs. *Sustainability, 14*(3), 1883. https://doi.org/10.3390/su14031883

Darling-Hammond, L. (2004). The color line in American education: Race, resources, and student achievement. Du Bois review: Social science research on. *Race, 1*(02). https://doi.org/10.1017/s1742058x0404202x

Davoodi, A. (2024). EQUAL AI: A framework for enhancing equity, quality, understanding and accessibility in Liberal arts through AI for multilingual learners. *Language, Technology, and Social Media, 2*(2), 178–203. https://doi.org/10.70211/ltsm.v2i2.139

Eden, C., Chisom, O., & Adeniyi, I. (2024). Integrating AI in education: Opportunities, challenges, and ethical considerations. *Magna Scientia Advanced Research and Reviews, 10*(2), 006–013. https://doi.org/10.30574/msarr.2024.10.2.0039

Edeni, C., Adeniyi, I., & Adeleye, O. (2024). The role of AI-enhanced tools in overcoming socioeconomic barriers in education: A conceptual analysis. *World Journal of Advanced Research and Reviews, 21*(3), 944–951. https://doi.org/10.30574/wjarr.2024.21.3.0780

Fan, Z., Wen, S., & Yan, Z. (2023). Deep learning and artificial intelligence in sustainability: A review of SDGs, renewable energy, and environmental health. *Sustainability, 15*(18), 13493. https://doi.org/10.3390/su151813493

Fang, M. L., Canham, S. L., Sixsmith, J., Battersby, L., Wada, M., & Sixsmith, A. (2018). Exploring privilege in the digital divide: Implications for theory, policy, and practice. *The Gerontologist, 59*(1). https://doi.org/10.1093/geront/gny037

Farooq, F., Rathore, F. A., & Mansoor, S. N. (2020). Challenges of online medical education in Pakistan during COVID-19 pandemic. *Journal of the College of Physicians and Surgeons Pakistan, 30*(6), 67–69. https://doi.org/10.29271/jcpsp.2020.supp1.s67

Filho, W. L., Filho, W. L., Sierra, J., Sierra, J., Price, E., Eustachio, J. H. P. P., Novikau, A., Kirrane, M., Kirrane, M., Dinis, M. A. P., Dinis, M. A. P., & Salvia, A. L. (2024). The role of universities in accelerating the sustainable development goals in Europe. *Scientific Reports, 14*(1). https://doi.org/10.1038/s41598-024-65820-9

García Peñalvo, F. J., Alier, M., Pereira, J., & Casany, M. J. (2024). Safe, transparent, and ethical artificial intelligence. *IJERI: International Journal of Educational Research and Innovation, 22*, 1–21. https://doi.org/10.46661/ijeri.11036

Gaur, A. S., Kumar, R., & Sharan, H. O. (2024). AI in education. *igi global*, 39–54. https://doi.org/10.4018/979-8-3693-2964-1.ch003

George, E. S. (2006). Positioning higher education for the knowledge based economy. *Higher Education, 52*(4), 589–610. https://doi.org/10.1007/s10734-005-0955-0

Gherheş, V., & Obrad, C. (2018). Technical and humanities students' perspectives on the development and sustainability of artificial intelligence (AI). *Sustainability, 10*(9), 3066. https://doi.org/10.3390/su10093066

Giangrande, N., East, M., Penha-Lopes, G., White, R. M., Clarke, T., Saloff Coste, M., & Jackson, R. (2019). A competency framework to assess and activate education for sustainable development: Addressing the UN sustainable development goals 4.7 challenge. *Sustainability, 11*(10), 2832. https://doi.org/10.3390/su11102832

Gilbert, C., & Gilbert, M. A. (2024). The convergence of artificial intelligence and privacy: Navigating innovation with ethical considerations. *International Journal of Scientific Research and Modern Technology (IJSRMT), 3*(9), 9–17. https://doi.org/10.38124/ijsrmt.v3i9.45

Gonzales, A. L., Lynch, T., & Mccrory Calarco, J. (2018). Technology problems and student achievement gaps: A validation and extension of the technology maintenance construct. *Communication Research, 47*(5), 750–770. https://doi.org/10.1177/0093650218796366

Grubaugh, S., Deever, D., & Levitt, G. (2023). Harnessing AI to power constructivist learning: An evolution in educational methodologies. *EIKI Journal of Effective Teaching Methods, 1*(3). https://doi.org/10.59652/jetm.v1i3.43

Hadjeris, F. (2021). Revisiting sustainable development goal 4 in the context of COVID-19 pandemic: A case study of online teaching in Algerian higher education institutions. *Human Behavior and Emerging Technologies, 3*(1), 160–168. https://doi.org/10.1002/hbe2.245

Han, S. P., & Kumwenda, B. (2024). Bridging the digital divide: Promoting equal access to online learning for health professions in an unequal world. *Medical Education, 59*(1). https://doi.org/10.1111/medu.15455

Hanemann, U. (2019). Examining the application of the lifelong learning principle to the literacy target in the fourth sustainable development goal (SDG 4). *International Review of Education, 65*(2), 251–275. https://doi.org/10.1007/s11159-019-09771-8

Heleta, S., & Bagus, T. (2020). Sustainable development goals and higher education: Leaving many behind. *Higher Education, 81*(1), 163–177. https://doi.org/10.1007/s10734-020-00573-8

Huang, L. (2023). Ethics of artificial intelligence in education: Student privacy and data protection. *Science Insights Education Frontiers, 16*(2), 2577–2587. https://doi.org/10.15354/sief.23.re202

Hutson, J., Leary, M., Arnone, K., Lively, J., Jeevanjee, T., Vosevich, K., Weber, J., Plate, D., Graaf, V. V., Edele, S., Weir, G., & Carnes, G. (2022). Artificial intelligence and the disruption of higher education: Strategies for integrations across disciplines. *Creative Education, 13*(12), 3953–3980. https://doi.org/10.4236/ce.2022.1312253

Ismail, I. A. (2024). Protecting privacy in AI-enhanced education. *igi global.*, 117–142. https://doi.org/10.4018/979-8-3693-0884-4.ch006

Janoušková, S., Hák, T., & Moldan, B. (2018). Global SDGs assessments: Helping or confusing indicators? *Sustainability, 10*(5), 1540. https://doi.org/10.3390/su10051540

Kara, A. (2021). COVID-19 pandemic and possible trends into the future of higher education: A review. *Journal of Education and Educational Development, 8*(1). https://doi.org/10.22555/joeed.v8i1.183

Khan, S., Mazhar, T., Shahzad, T., Khan, M. A., Rehman, A. U., Saeed, M. M., & Hamam, H. (2025). Harnessing AI for sustainable higher education: Ethical considerations, operational efficiency, and future directions. *Discover Sustainability, 6*(1). https://doi.org/10.1007/s43621-025-00809-6

Lazarus, M. D., Douglas, P., Truong, M., & Selwyn, N. (2022). Artificial intelligence and clinical anatomical education: Promises and perils. *Anatomical Sciences Education, 17*(2), 249–262. https://doi.org/10.1002/ase.2221

Le Blanc, D. (2015). Towards integration at last? The sustainable development goals as a network of targets. *Sustainable Development, 23*(3), 176–187. https://doi.org/10.1002/sd.1582

Leal Filho, W., Wu, Y. C. J., Brandli, L. L., Avila, L. V., Azeiteiro, U. M., Caeiro, S., & Madruga, L. R. D. R. G. (2017). Identifying and overcoming obstacles to the implementation of sustainable development at universities. *Journal of Integrative Environmental Sciences, 14*(1), 93–108. https://doi.org/10.1080/1943815x.2017.1362007

Leffia, A., Kiboyi, A., & Terizla, R. (2024). Implementation of the use of artificial intelligence in higher education. *Blockchain Frontier Technology, 3*(2), 150–153. https://doi.org/10.34306/bfront.v3i2.478

Li, D., & Hu, S. (2021). How does technological innovation mediate the relationship between environmental regulation and high-quality economic development? *Empirical Evidence from China. Sustainability, 13*(4), 2231. https://doi.org/10.3390/su13042231

López-Chila, R., Llerena-Izquierdo, J., Cueva-Estrada, J., & Sumba-Nacipucha, N. (2023). Artificial intelligence in higher education: An analysis of existing bibliometrics. *Education Sciences, 14*(1), 47. https://doi.org/10.3390/educsci14010047

Ma, X., & Jiang, C. (2023). On the ethical risks of artificial intelligence applications in education and its avoidance strategies. *Journal of Education, Humanities and Social Sciences, 14*, 354–359. https://doi.org/10.54097/ehss.v14i.8868

Magrill, J., & Magrill, B. (2024). Preparing educators and students at higher education institutions for an AI-driven world. *Teaching and Learning Inquiry, 12*. https://doi.org/10.20343/teachlearninqu.12.16

Mah, C., Walker, H., Levine, S., Phalen, L., Beck, S. W., & Pittman, J. (2024). Beyond CheatBots: Examining tensions in teachers' and students' perceptions of cheating and learning with ChatGPT. *Education Sciences, 14*(5), 500. https://doi.org/10.3390/educsci14050500

Maity, S., & Deroy, A. (2024). Generative AI and its impact on personalized intelligent tutoring systems. *center for open science*. https://doi.org/10.35542/osf.io/kawr5

Malerbi, F. K., Villanueva, C., Nakayama, L. F., Zago Ribeiro, L., Gayle Dychiao, R., Regatieri, C. V., & Celi, L. A. (2023). Digital education for the deployment of artificial intelligence in health care. *Journal of Medical Internet Research, 25*, e43333. https://doi.org/10.2196/43333

Mccowan, T. (2015). Three dimensions of equity of access to higher education. *Compare: A Journal of Comparative and International Education, 46*(4), 645–665. https://doi.org/10.1080/03057925.2015.1043237

Mccowan, T. (2023). The crosscutting impact of higher education on the sustainable development goals. *International Journal of Educational Development, 103*, 102945. https://doi.org/10.1016/j.ijedudev.2023.102945

Mcgowan, P. J. K., Stewart, G. B., Long, G., & Grainger, M. J. (2018). An imperfect vision of indivisibility in the sustainable development goals. *Nature Sustainability, 2*(1), 43–45. https://doi.org/10.1038/s41893-018-0190-1

Mikhaylov, S. J., Campion, A., & Esteve, M. (2018). Artificial intelligence for the public sector: Opportunities and challenges of cross-sector collaboration. *Philosophical Transactions of the Royal Society A: Mathematical, Physical and Engineering Sciences, 376*(2128), 20170357. https://doi.org/10.1098/rsta.2017.0357

Mollura, D. J., Scheel, J. R., Battino, G., Culp, M. P., Schweitzer, A., Elahi, A., Pollack, E., Mango, V. L., & Dako, F. (2020). Artificial intelligence in low- and middle-income countries: Innovating Global Health radiology. *Radiology, 297*(3), 513–520. https://doi.org/10.1148/radiol.2020201434

Morandín-Ahuerma, F. (2024). A critical analysis of the European Union's considerations on the ethical use of artificial intelligence in education. *center for open science*. https://doi.org/10.31219/osf.io/escqj

Murdoch, B. (2021). Privacy and artificial intelligence: Challenges for protecting health information in a new era. *BMC Medical Ethics, 22*(1). https://doi.org/10.1186/s12910-021-00687-3

Nedungadi, P., Raman, R., & Tang, K. Y. (2024). The transformative power of generative artificial intelligence for achieving the sustainable development goal of quality education. *Sustainability, 16*(22), 9779. https://doi.org/10.3390/su16229779

Ochuba, N., Adewumi, A., & Olutimehin, D. (2024). The role of AI in financial market development: enhancing efficiency and accessibility in emerging economies. *Finance & Accounting Research Journal, 6*(3), 421–436. https://doi.org/10.51594/farj.v6i3.969

Porto, M. (2009). Culturally responsive L2 education: An awareness-raising proposal. *ELT Journal, 64*(1), 45–53. https://doi.org/10.1093/elt/ccp021

Price, E. A. C., Mori, K., Preist, C., Baughan, P., White, R. M., Hayles, C. S., Gough, G., & Longhurst, J. (2021). Supporting the role of universities in leading individual and societal transformation through education for sustainable development. *Discover Sustainability, 2*(1). https://doi.org/10.1007/s43621-021-00058-3

Purcell, W. M., Henriksen, H., & Spengler, J. D. (2019). Universities as the engine of transformational sustainability toward delivering the sustainable development goals. *International Journal of Sustainability in Higher Education, 20*(8), 1343–1357. https://doi.org/10.1108/ijshe-02-2019-0103

Qian, C. (2023). Research on human-centered Design in College Music Education to improve student experience of artificial intelligence-based information systems. *Journal of Information Systems Engineering and Management, 8*(3), 23761. https://doi.org/10.55267/iadt.07.13854

Ramirez, F. O., & Tiplic, D. (2013). In pursuit of excellence? Discursive patterns in European higher education research. *Higher Education, 67*(4), 439–455. https://doi.org/10.1007/s10734-013-9681-1

Rio, C. J., & Saligan, L. N. (2023). Understanding physical activity from a cultural-contextual lens. *Frontiers in Public Health, 11*. https://doi.org/10.3389/fpubh.2023.1223919

Rizvi, M. (2023). Investigating AI-powered tutoring systems that adapt to individual student needs, providing personalized guidance and assessments. *The Eurasia Proceedings of Educational and Social Sciences, 31*, 67–73. https://doi.org/10.55549/epess.1381518

Saborío-Taylor, S., & Rojas-Ramírez, F. (2024). Universal design for learning and artificial intelligence in the digital era: Fostering inclusion and autonomous learning. *International Journal of Professional Development, Learners and Learning, 6*(2), ep2408. https://doi.org/10.30935/ijpdll/14694

Sanusi, I. T., & Olaleye, S. A. (2022). An insight into cultural competence and ethics in K-12 artificial intelligence education, *12*, 790–794. https://doi.org/10.1109/educon52537.2022.9766818

Scillitoe, J. L., Poonamallee, L., & Joy, S. (2018). Balancing market versus social strategic orientations in socio-tech ventures as part of the technology innovation adoption process—Examples from the global healthcare sector. *Journal of Social Entrepreneurship, 9*(3), 257–287. https://doi.org/10.1080/19420676.2018.1498378

Shulla, K., Voigt, B.-F., Scandone, G., Martinez, E., Nelkovski, F., Salehi, P., & Cibian, S. (2021). Effects of COVID-19 on the sustainable development goals (SDGs). *Discover Sustainability, 2*(1). https://doi.org/10.1007/s43621-021-00026-x

Singh, A., & Blessinger, P. (2023). Examining the role and challenges of sustainable development goals for the universities in The United Arab Emirates. *Sustainability, 15*(20), 15123. https://doi.org/10.3390/su152015123

Slimi, Z., & Villarejo Carballido, B. (2023). Navigating the ethical challenges of artificial intelligence in higher education: An analysis of seven global AI ethics policies. *TEM Journal*, 590–602. https://doi.org/10.18421/tem122-02

Song, J., & Jang, C. (2023). Unpacking the sustainable development goals (SDGs) interlinkages: A semantic network analysis of the SDGs targets. *Sustainable Development, 31*(4), 2784–2796. https://doi.org/10.1002/sd.2547

Sporn, B., & Godonoga, A. (2024). Higher education institutions as change agents in society: Perspectives on adaptation and impact. *European Journal of Higher Education, 14*(sup1), 1–9. https://doi.org/10.1080/21568235.2024.2412764

Trott, C. D., Sample Mcmeeking, L. B., & Weinberg, A. E. (2018). Prefiguring sustainability through participatory action research experiences for undergraduates: Reflections and recommendations for student development. *Sustainability, 10*(9), 3332. https://doi.org/10.3390/su10093332

Unterhalter, E. (2014). Measuring education for the millennium development goals: Reflections on targets, indicators, and a Post-2015 framework. *Journal of Human Development and Capabilities, 15*(2–3), 176–187. https://doi.org/10.1080/19452829.2014.880673

Van Zanten, J. A., & Van Tulder, R. (2021). Improving companies' impacts on sustainable development: A nexus approach to the SDGS. *Business Strategy and the Environment, 30*(8), 3703–3720. https://doi.org/10.1002/bse.2835

Verboven, H., & Vanherck, L. (2016). Sustainability management of SMEs and the UN sustainable development goals. *Uwf UmweltWirtschaftsForum, 24*(2–3), 165–178. https://doi.org/10.1007/s00550-016-0407-6

Volk, S. C., Vogler, D., Fürst, S., Schäfer, M. S., & Sörensen, I. (2023). Role conceptions of university communicators: A segmentation analysis of communication practitioners in higher education institutions. *Public Relations Review, 49*(4), 102339. https://doi.org/10.1016/j.pubrev.2023.102339

Wang, C. L., & Rafiq, M. (2012). Ambidextrous organizational culture, contextual ambidexterity and new product innovation: A comparative study of UK and Chinese high-tech firms. *British Journal of Management, 25*(1), 58–76. https://doi.org/10.1111/j.1467-8551.2012.00832.x

Weiland, S., Lederer, M., Hickmann, T., Marquardt, J., & Schwindenhammer, S. (2021). The 2030 agenda for sustainable development: Transformative change through the sustainable development goals? *Politics and Governance, 9*(1), 90–95. https://doi.org/10.17645/pag.v9i1.4191

Wu, L., Yan, K., & Zhang, Y. (2020). Higher education expansion and inequality in educational opportunities in China. *Higher Education, 80*(3), 549–570. https://doi.org/10.1007/s10734-020-00498-2

Yeter, I. H., Sturgess, J. B., & Yang, W. (2024). Global initiatives and challenges in integrating artificial intelligence literacy in elementary education: Mapping policies and empirical literature. *Future in Educational Research, 2*(4), 382–402. https://doi.org/10.1002/fer3.59

Yew, R.-J., Venkatasubramanian, S., & Qin, L. (2024). You still see me: How data protection supports the architecture of AI surveillance. *Proceedings of the AAAI/ACM Conference on AI, Ethics, and Society, 7*, 1709–1722. https://doi.org/10.1609/aies.v7i1.31759

Yılmaz, Ö. (2024). Personalised learning and artificial intelligence in science education: Current state and future perspectives. *Educational Technology Quarterly, 2024*(3), 255–274. https://doi.org/10.55056/etq.744

Zekaj, R. (2023). AI language models as educational allies: Enhancing instructional support in higher education. *International Journal of Learning, Teaching and Educational Research, 22*(8), 120–134. https://doi.org/10.26803/ijlter.22.8.7

Zhou, X., & Schofield, L. (2024). Developing a conceptual framework for artificial intelligence (AI) literacy in higher education. *Journal of Learning Development in Higher Education, 31*. https://doi.org/10.47408/jldhe.vi31.1354

Evolution of AI in Higher Education

Renjith V. Ravi and Aftab Ara

Abstract Over the past ten years, integration of AI in higher education has changed from application in sub-conventional modes to enhancement of learning experiences. Briefly, this review assesses the development of AI technologies and their utilization in higher education, particularly concerning achievement of SDG 4: Quality Education through potential AI. As it enables personalized learning, administrative tasks, and provides intelligent tutoring systems, AI encourages quality education for all among other principles. The paper also tackles the operational and ethical issues that arise out of the use of AI technologies in learning settings such as low and negative bias/ prejudice, privacy of data and the digital gap. At the same time and in correlation to the topic being discussed in this review, the future tendencies of education revolutionized by AI technologies emphasize such cutting-edge forms as adaptive learning, AI assessments, and virtual learning. Finally, the review evaluates the changing trends in AI- enabled education towards support of lifelong learning and sustenance of lifelong education by providing ideas on how AI will help realize the world education goals.

Keywords Artificial intelligence (AI) · Higher education · SDG 4 · Quality education · Personalized learning · Intelligent tutoring systems · Adaptive learning · Educational technology · Lifelong learning · Digital divide

1 Introduction

Artificial Intelligence (AI) is changing the power of the society and education sector is the most affected as AI is being transformed. Among these, higher education has been becoming more and more involved in using AI technologies for improving

R. V. Ravi (✉)
School of Applied Sciences Engineering and Technology, Rashtriya Raksha University, Gandhinagar, India

R. V. Ravi · A. Ara
College of Business Administration, University of Hail, Hail, Saudi Arabia
e-mail: ara.aftab@gmail.com

A. Ara (ed.), *Role of Artificial Intelligence in Higher Education*,
https://doi.org/10.1007/978-981-95-5350-1_2

learning experience, optimizing administrative process and broadening accessibility (Bhatnagar, 2024). Educational institutions across the world are faced with a challenge to deliver quality education and high standards at scale, and AI has so far proven to be a suitable tool to overcome these complex problems.

While AI is being introduced in education as an administrative task automate, its deployment extends beyond that, into customized learning, adaptive assessment, intelligent tutoring, as well as for predictive analytics (Adepoju, 2025). The potential offered by these new applications is one of having education that is not only more effective but also more inclusive and hence equal opportunity for all to have access to high quality educational resources.

As for the AI technology, it is changing which demands us to research and analyses it further to utilize all its benefits and, at the same time, reduce risks. The pace of AI developments is becoming very fast, such that education providers, policymakers and technology developers needed to work together to ensure navigation of the nuances and implications of its deployment in higher education settings [Butson & Spronken-Smith, 2024). Proactively addressing these issues guarantees that the institutions' AI technologies are pedagogically aligned and meet ethical standards.

1.1 Overview of AI in Education

In education, AI consists of many computational technologies and methods that aim to replicate the human style of intelligence such as learning, reasoning, problem solving and adaptability (Gao et al., 2023). Educational technology (EdTech) traditionally was centred on basic digital tools for dissemination of content and instructional support in a classroom. While the tools had the potential to be AI based, today AI based systems are advanced and are doing their own analysis, offering personalized content and interacting with students on a real time basis.

The popularity of AI driven systems is growing quite a lot because it offers tailor made educational experience for each learner. AI based personalized learning environments are focused on developing multiple learner—specific factors including previous knowledge, learning pace, and learning preferences and converged towards learner's unique educational pathway (Kinshuk, 2016). In addition, these systems can always adapt to students' changing needs and provide timely interventions and information for better learning outcome.

A unique part of AI contributions into the educational sector is personalization, but other than this, it helps to increase efficiency in administrative and operational tasks within educational organizations. AI facilitates automating grading and offering immediate feedback on students work for educators, and handling large institutional databases which takes the burden of administrative tasks out of educators system and helps them spend more time providing instruction and interacting with students (Zhou, 2023).

Some examples of current AI applications in higher education are summarized as below in Table 1.

Table 1 AI applications in higher education

AI Application	Functionality	Example
Personalized learning	Modifying the content of education and education pathway as per the needs of individual students	Smart Sparrow, Knewton
Intelligent tutoring systems	Their role is providing instructional support and feedback in a personalized manner	Carnegie learning, Thinkster math
Adaptive assessments	Adapts difficulty on demand to the learners' responses	GMAT, GRE adaptive tests
Administrative automation	The problem statement is how to streamline operations, such as grading, scheduling, and student inquiries, so as to reduce the load on instructors	Gradescope, AdmitHub
Predictive analytics	It aims at identifying at risk students and predicting the academic performance	Civitas learning, blackboard analytics

1.2 Relevance of AI to SDG 4: Quality Education

The Sustainable Development Goal 4 (SDG 4) as specified by the United Nations is to 'ensure inclusive and equitable quality education and promote lifelong learning opportunities for all' (Chaney, 2022). To achieve this goal, it requires innovative solutions that deal with the classical problems of education: isolation (geographical), lack of resources and different learning abilities. SDG 4 and the opportunities related to with AI perfectly align together.

At first, it does the thing of inclusive education by making the learning environment personalized and adaptive in such a way that the student body is served. With AI backed tools now, students who have been traditionally disadvantaged at getting education on account of things like disabilities, language differences, inability to get a competent teacher can now be supported with personalized coaching (Fabiano et al., 2024). In addition to that, AI systems can transliterate multiple languages at any time turning linguistic barriers into seeming walls.

Second, AI does that accomplish educational equity by democratizing access to great quality resources and specialist instruction. Given the slow growth in these institutions' staff, adopting AI driven platforms can be of great use in providing students access to educational content that was never available to them, thereby reducing the global educational divide. It gives those who opt for it the tailored educational experience they can ever receive through personal tutoring that was once the preserve of the privileged.

Lastly, AI supports lifelong learning initiatives crucial for continuous skill development in the evolving job market (Yadav, 2024). That is the unique offering of AI powered learning platforms that provides flexible, self-paced learning and a viable option for lifelong learning to adults who can't spend the time needed due to other commitments like employment. For this reason, AI assists in the creation of a workforce that possesses the required skills to work amid an ever change technological world.

Finally, the application of AI in the field of higher education signifies massive transformation of the overall educational practice that leads toward a more personalized, equal, and efficient one. The implementation of the objectives within the scope of SDG 4 has the power to substantially restructure how education is delivered and accessible through the use of AI with the intention of reaching globally recognized standards of quality education.

2 Evolution of AI in Higher Education

The growth of Artificial Intelligence (AI) has gradually risen as a prominent matter of change within higher education. In the recent decades, AI has indeed seeped into educational settings and changed the ways of pedagogy, administrative functions, and student involvement (Chouzouris, 2024). The development from simple, rule-based systems to complex machine learning algorithms that are able to make the decision by itself is a substantial change in how educational entities handle teaching, learning, and management. In this section we will look at the historical trajectory and main landmarks of the recent history of artificial intelligence among the institutions of higher learning.

2.1 Historical Development of AI in Education

The history of AI in education starts from the middle of the twentieth century, when it was in its primitive form in the form of rudimentary computer-based training (CBT) programs and simple tutoring programs aimed at delivery of pre-defined instructional content (Tariq, 2024). The first generation of Intelligent Tutoring Systems (ITS) are based on early AI researchers' vision to build intelligent systems that can simulate a human tutor. These systems utilized rule-based logic and some basic heuristics to steer students through educational contents offering immediate response as per preset scenarios.

With tremendous technological advancement that is now taking place, AI technologies became increasingly sophisticated by the late 1980s and early 1900s. Machine learning algorithms, natural language processing and data mining started to become prominent such that such that they provided more sophisticated, interactive and personalized learning experience. One of the pioneering intelligent tutoring programs, SCHOLAR, could go far as systems that would interact with students in text-based dialogues.

As the internet continued to gain increased adaptation in the following years (2000s), the integration of AI in higher education continued to grow (Bardhan & Kandarp, 2025). As such, online education platform started using AI driven recommendation engines to customize course materials to the needs of the individual learner. AI integration extended beyond teaching to include administrative domains for enhancing process such as student enrollment, grading and academic advising.

2.2 Milestones in AI Integration in Higher Education Institutions

During the last few decades, many distinguished milestones have been left with regards to integrating AI to higher education (Temper, 2025). Along with being technological milestones, they represent a change in institutional attitudes towards embracing of digital transformation. The following table in Table 2 gives a brief on critical milestones and its impact of AI into higher education:

In each of these milestones an important transition in the application of AI is emphasized for using the technology in education, from basic rules-based programming to using algorithms that can adapt dynamically to the needs and behavior of the learner. In 2010's, AI based adaptive platforms, such as Knewton (Conklin, 2016) and Smart Sparrow (Sparrow, 2025) implemented a milestone on personalization and effectiveness. In these platforms, data analytics in real time were deployed to adjust instruction in order to supply the right resources and interventions at the exact moment when they were needed most.

Over the past few years, AI technologies have been increased with the COVID pandemic, inducing a proliferation of virtual learning environments improved by

Table 2 Critical milestones and its impact of AI into higher education

Year/Period	Milestone	Impact on Higher Education
1960–1970	Development of basic computer-assisted instruction (CAI) systems	Investigating the use of AI with respect to education methods
1980–1990	Emergence of intelligent tutoring systems (e.g., SCHOLAR)	Interactive dialogue-based learning, personalized instructions and learning
2000–2010	The introduction of learning management systems (LMS) in the educational sector that is enhanced by AI algorithms	Personalised and online education, wider access, automated assessments
2010–2020	Proliferation of adaptive learning platforms (e.g., Knewton, smart Sparrow)	More personalized and real time of personalizing the course to the student learning styles and pacing
2020-present	Integration of generative AI (e.g., ChatGPT, GPT models) and virtual learning environments	It offers enhanced interactive, scalable and virtual/hybrid learning models support

generative AI technologies, like OpenAI's GPT models (Singh & Singh, 2023). Through these advancements, it was possible for institutions to deal with the sudden change to remote education efficiently and provide scalable solutions for tutoring, assessment and administrative functions. During this period, it demonstrated how AI can keep educational continuity and quality going while hunting unusual conditions in higher education related to a new level of resilience and adaptability.

The path of AI's assimilation into the realm of higher education seems destined to continue, as the first steps are taken in conjunction with continued advances in the power of AI and a desire for the education to embrace everyone. Worldwide, institutions are means to apply advanced analytics, ethics AI frameworks and collaborative AI Human learning strategies that enhance learning outcomes and access equitable education opportunity.

Briefly, evolution of AI within the higher education environment can be described as a continuous process from simple automation of instructional tasks to complex systems with AI capable of generating very sophisticated and dynamic responses. This represents the evolution of technological growth as well as the evolution of the educational philosophy that pointed at AI as a key factor that will allow us to create responsive and inclusive and effective educational environments.

3 AI Applications in Higher Education

Artificial intelligence (AI) has evolved many of the aspects of higher education, including administrative efficiency, instructional strategies, to student involvement (Kurtzke, 2024). Over the past years, globally, there has been a growing combination of AI-driven solutions to create more dynamic, responsive and customizable learning environment. Along these lines, these advances along with the personalized learning, intelligent tutoring systems, administrative effectiveness, and automated assessment methods show how artificial intelligence contributes to the expansion and effectiveness of higher education.

Giving virtual tutoring solutions, automating evaluation, making instructional materials more specific to the student, or simplifying admin work—these are exactly what the organizations can rely on the help of the AI applications. Higher education institutions show so much in these sectors they guarantee better quality of instruction by incorporating the artificial intelligence and at the same time have greatly increased their operational efficiency and responsiveness of operations. The application of AI in higher education is depicted in Fig. 1.

3.1 Personalized Learning and AI

The instructional experiences are adjusted by artificial intelligence driven personalised learning to support individual student's needs, interests, aptitudes and rate

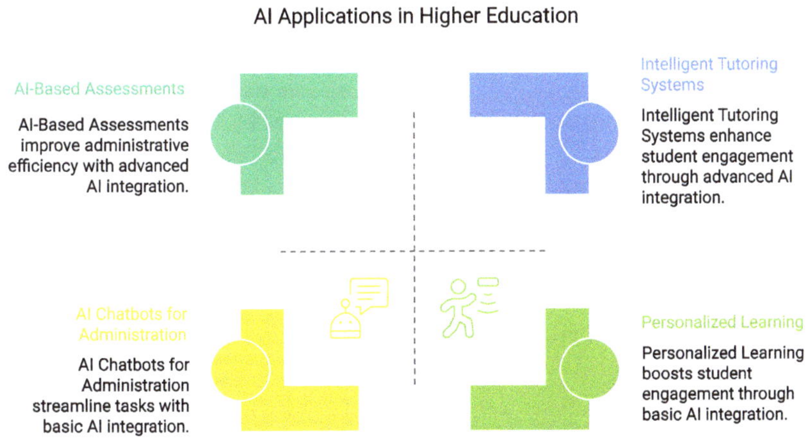

Fig. 1 Application of AI in Higher Education

of learning (Posavec, 2025). Artificial intelligence systems acquire and assess huge amounts of data about a student's performance, preferences and interactions so as to build personalized learning paths for them. This customised method ensures that students closely interact with the learning resources, thereby increasing the learning and the retention rates.

Personalised learning includes adaptive learning technology and it is one that dynamically alters the complexity of the instructional materials based on real time evaluations of student competency. For example, Smart Sparrow or Knewton is an AI system that provides tailored modules to students, the tailored modules being a module that reacts to how the knowledge of the student is changing and to the changes in his skills, thereby serving to improve the learning process.

Artificial intelligence, on the other hand, also helps pupils at risk be proactively intervened as it sees early indicators of the academic difficulty. By means of predictive analytics, institutions are able to be in the lead in the recognition of the utilization of trends that suggest possible problems, allowing teachers to immediately take measures in the shape of custom coaching or additional tools.

3.2 Intelligent Tutoring Systems

Artificial intelligence also makes another major development possible in the way of intelligent tutoring systems (ITS), which provide responding to comments and guidance similar to a human instructor (Ferster, 2022). But these systems combine cognitive modelling, machine learning and natural language processing (NLP) in order to be able to replicate human tutoring interactions. Eminent ITS systems like the Carnegie Learning MATHia and ALEKS (Assessment and Learning in Knowledge

Spaces) address individual misunderstandings and knowledge gaps one at a time through focused and instantaneous feedback, which highly improve student learning.

ITS technology also provides very helpful scalable tutoring solutions in distant learning situations or large-scale educational settings, when it is not possible to have individual human tutoring. Furthermore, its inclusion in STEM subjects has improved student performance significantly, which includes speculation that the potential for use in many other disciplines such as humanities and social sciences is great.

3.3 AI in Administrative and Institutional Processes

In addition to the applications with regard to the classroom, artificial intelligence plays a significant role in raising administrative efficiency of high educational organizations (Irshad, 2024). Artificial intelligence simplifies administrative responsibilities such as enrolment management, resource allocation, scheduling, and manages student in case of resource allocations. Automation of mundane, repetitive tasks by AI enables administrative personnel to focus more directly on the strategic goals and student center activities leading to an increased institutional efficiency.

A well encountered example would be the use of artificial intelligence powered chatbots such as Georgia State University's Pounce chatbot that helps manage enquiries regarding enrolment, financial aid and other general student services, reducing response times and administrative burden by a great extent. AI analytics also precisely and with an incredible speed predict enrolment patterns from facility use, personnel needs and institutional planning and resource allocation.

3.4 Ai-Based Assessments and Grading

Because AI based assessment solutions automate grading process, the feedback given is fast, accurate, detailed and consistent (Tobler, 2023). Such AI systems as Turnitin and Gradescope allow quickly processing of assignments, tests and essays via advanced machine learning techniques and natural language processing. Because these instruments provide quick and comprehensive analysis of student performance, they can be used to create a real time feedback loop around continual growth.

For these reasons, the utilization of AI driven testing is able to reduce grading time, ensure uniformity and fairness, leaving teachers with more time to concentrate on student mentoring and curriculum making. Also, AI based evaluations guarantee that the same criteria is applied to all student entries which ensures objective grading and ensures fair educational results.

4 AI and SDG 4: Quality Education

In fact, in the educational sector, Artificial Intelligence (AI) has been found to be very useful and is in complete alignment with Sustainable Development Goal 4 (SDG 4), i.e. inclusive and equitable quality education and promoting lifelong learning opportunities for all (Hales & Birdthistle, 2022). AI technology provides creative ways to solve the problems of access, personalization, and effectiveness of education, which directly contributes to the achievement of the goals set out in SDG 4. AI is the pivotal instrument that addresses some of the key educational challenges, such as resource constraint, gaps in educational quality, and barriers to lifelong learning in order to fulfill the promises associated with the global educational agenda.

4.1 AI's Role in Achieving SDG 4

SDG 4 refers to universal, equitable quality education and lifelong learning for everyone. It aims at ensuring youth and adult literacy as well as the removal of gender disparities in education (Mariotti, 2019). Enabling universal access to quality technical, vocational training, and lifelong learning is a part of this goal, while broadening the access to information and communication technology in classrooms helps in bridging the digital divide. AI helps to meet SDG 4 by tackling persisting education inequalities and improving educational outcomes through adaptive and customized approaches. Adaptive learning systems and intelligent tutoring systems (ITS) are some example of AI driven technologies which can dynamically tailor to needs of individual student, schemes the content and pace of instruction thus can bring down the learning outcomes. These systems permit the learning of classes of learners with different background profiles in a personalized manner, while removing the gaps induced by the traditional education constraints.

Additionally, AI based analytics goes a long way in spotting at-risk students early and thus rescuing from high dropout rates because of such timely interventions. For instance, these predictive analytics give educational support to teachers who can act based on these insights, making informed decisions and based on customized educational support. Additionally, the use of AI can help to automatically conduct administrative tasks so that educators would have more time to focus on their main tasks, namely direct instructional work, student engagement, etc.

4.2 AI's Contribution to Inclusive and Equitable Education

Democratization of education through AI has the capacity to spread and include more people into education. With the use of artificial intelligence (AI) based translation service and natural language processing (NLP) tool, language barriers are broken down and quality education is facilitated to all across the world, despite of multi lingual and multi-cultural population (Lata, 2024). Speech recognition, text to speech and adaptive interfaces among other assistive AI technologies accommodate students with disabilities hence guaranteeing a level playing field for students in educational contexts.

Finally, the VLEs that are powered by AI also increase inclusivity by offering geographically isolated and socioeconomically disadvantaged students high quality educational resources. Thus, as AI is integrated in VLEs, learners wherever they may be, can access resources available yesterday to only some particular geographical regions or institutions.

4.3 *Promoting Lifelong Learning through AI*

In order to achieve SDG 4, lifelong learning is essential and with the use of AI technologies we now have more possibilities for continuous education than the classroom (Aggarwal, 2023). AI platforms make it possible for learners to have personalized learning pathways that responds to the changing skill requirements of learners and has to do with the advancement of careers and personal growth. Such platforms leverage powerful algorithms in order to sift relevant content, assessments and skill building activities from each individual's career target and the way they learn. Table 1 below provides a summary of specific AI technologies and their contributions to lifelong learning:

Table 3 summarizes some AI technologies that can bring human learning closer to life by facilitating lifelong learning.

One that is so important in a job market that's changing so quickly. Its purpose is to enable individuals to continuously update their skills, adjusting them to the

Table 3 Some AI technologies that can bring human learning closer to life by facilitating lifelong learning

AI Technology	Application in Lifelong Learning	Example Platforms
Adaptive learning systems	Focusing on education tailored to the user skills and their progress	Coursera, khan academy
Intelligent tutoring systems	Customized feedback and learning support	Carnegie learning, DreamBox
AI-based career guidance	Career recommendations and skills gap analysis	LinkedIn learning, EdCast
Virtual reality (VR) platforms	Immersive skill-based training and simulation	Oculus VR, ClassVR
AI-powered MOOCs	Scalable, flexible, and diverse course offerings	edX, FutureLearn

continuously changing professional requirements and to have a competitiveness in the labour market and ensure economic resilience. Additionally, AI facilitates self-directed learning as it gives the user a lot of freedom and autonomy of their journey of learning. Users can pick up relevant learning paths based on their interests or career aspirations and AI powered suggestions help to direct the users to suitable courses, skills and learning resources. This autonomy extends into life, giving it the ability to engage with educational processes sustainably into the future, and thus continue intellectual and professional development. Finally, incorporating AI in the educational field is in line to help bringing SDG 4 objectives to fulfillment through ensuring an inclusive, equitable, and quality education. AI contributes to a seamless access to quality education through the adaptive learning systems, intelligent tutoring, assistive technologies and lifelong learning platforms, facilitating a learning environment that can support various needs of learners. Though this is the case, stakeholders must always ensure that ethical considerations, as well as the digital divide, are always addressed so that AI technologies contribute positively to educational equity and sustainability internationally.

5 Challenges and Ethical Considerations

With that, the integration of Artificial Intelligence (AI) in higher education system has transformative opportunities on the one hand, but also brings in various challenges and ethical issues (Hamid, 2024). Deployment of AI in responsible buildings in educational settings necessitates critical thinking points such as avoidance of AI algorithm biases, robust data privacy, and resolve the digital divide so that they can reach to all learners equitably. It is therefore vital that we are able to understand and tackle these complexities that would allow us to use AI's full potential ethically and in a way that includes as many people as possible. A smart diagram showing all these challenges and ethical considerations are shown in Fig. 2.

5.1 Addressing Bias in AI Systems

Algorithmic bias is one of the biggest ethical problems that AI presents to higher education. Construction of AI can have bias in various stages, i.e., data collection, Algorithm design and system implementation (Pulivarthy & P. Whig, 2024). If not closely monitored, educational AI systems including those which are used in personalizing learning or assessments may happen to perpetuate existing inconveniences.

Most times, bias in AI systems is a result of training datasets that are not representative or that reflect historical inequities. Consider an example of a system for assessing students using AI where the training data is biased by majority groups,

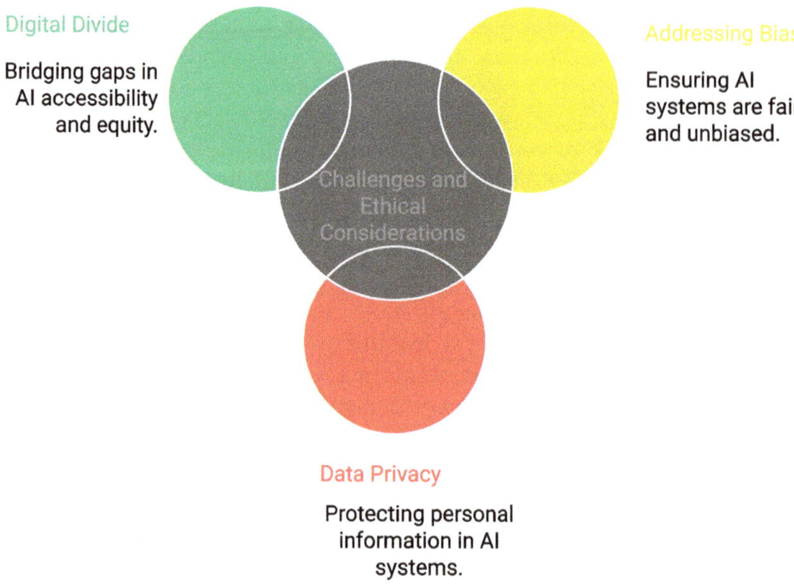

Fig. 2 Challenges and Ethical Considerations of AI in Education

which is an example of such disadvantage. Such a biased representation if operated upon can lead to unfair scoring, unfair recommendations and unfair resource allocation.

A thorough testing and validation of an algorithm with diverse datasets is needed in order to mitigate AI bias. However, institutions must institute mechanisms of oversight that involve different demographics of stakeholders for proactive identification and remediation of biases. To further enhance accountability, they can have regular audits and transparent reporting of the efforts. The Common Sources and Mitigation Strategies for AI Bias in Education is shown in Table 4.

Table 4 Common Sources and Mitigation Strategies for AI Bias in Education (Kamenskih, 2022) (Taylor, 2025)

Sources of Bias	Examples in Education	Mitigation Strategies
Data sampling bias	Underrepresentation of minority groups	Ensuring inclusive and representative datasets
Historical bias	Historical discrimination in assessment data	Regular audits and diversity-aware algorithms
Algorithmic bias	Favoring certain learner profiles	Transparency and interpretability in models
Confirmation bias	Reinforcing stereotypes in recommendations	Training and updation of the AI models continuously

The educational institutions must invest in training the stakeholders to familiarize them with the process of identifying biases in AI tools for enabling them to make an informed decision and use AI responsibly.

6 Emerging Trends in AI for Higher Education

Artificial Intelligence (AI), comprised as it is of machine learning ‘machines’, virtual assistants, smart sensors, chatbots, and various other AI applications have been reshaping the landscape of Higher Education altogether (Potluri & Kilaru, 2024). However, lately, advancements of AI technologies have paved the way for innovations in use of educational methodologies, administrative processes and the ways to engage students. AI continues to be an area where new trends point to taking on a lot of the challenges that will serve the educational space, from interactivity in virtual scenarios, personalization at scale, sustainability in education systems, among others. The section presents some emerging important current trends in the field of AI-based higher education that include adaptive learning platforms; AI-enabled virtual learning environments; and apps for educational sustainability.

6.1 Adaptive Learning Platforms

Of all the AI applications for higher education, adaptive learning platforms are among the most radical in terms of change (Jiao, 2024). These are platforms that utilize data analytics and sophisticated algorithms to tailor to the educational needs of the individual student with regard to their respective needs, pace of learning, style of learning and competency. By doing so, this personalized approach guarantees that neither learning will be hard nor easy and keep challenge at an optimum level as well as engagement level for each student.

One such benefit of adaptive learning platforms is that they gather and analyze learner data to forecast learning result and modify the content in real time. For instance, Knewton and Smart Sparrow based platforms monitor in real time gaps in student’s knowledge, and offer them personalized resources or remedial intervention. In this way, the odds of student disengagement or academic frustration are minimized and greatly help to bring up educational outcomes.

Moreover, adaptive platforms create more inclusive learning environment as it addresses different learning styles and needs and can be more helpful for learners with disability or from marginalized background. Adaptive learning systems enable democratisation of quality education, and share close alignment with global initiatives such as Sustainable Development Goal 4 (SDG 4).

6.2 AI-Powered Virtual Learning Environments

Nevertheless, global events have catalysed digital education transformation to such a degree that AI powered virtual learning environments are starting to be recognised for their value (Buşu, 2024). AI help replicate dynamic and interaction experience of a normal classroom in virtual learning space (VLE). This brings a lot of improvement to the online learning.

The use of intelligent tools such as virtual assistants, automated grading system, and also immersive simulations can empower AI in powering VLEs and bring an interactive and motivating learning environment. For example, virtual assistants that are embedded in platforms such as Blackboard Learn and Canvas can help students organize tasks (e.g., set reminders, provide real time feedback to students), answer most frequently asked questions, etc. First, these automated interactions that go on between the instructor and learner reduce the administrative burden of the educator such that they can focus on pedagogic tasks.

In addition to this, AI facilitated VLE's can use the augmented reality (AR) and virtual reality (VR) technologies to enhance educational experiences. In subjects such as medicine, engineering or natural sciences, students can adopt highly realistic simulations and practical scenarios to comprehend the topics with experiential learning, thus improving the understanding considerably. Besides, AI integration in VLE makes it possible to make the interactions between students and teachers better, allowing educators to keep tabs on the student engagement and performance on a real time basis and take counter actions before the situation worsens.

6.3 AI for Educational Sustainability

Therefore, the term educational sustainability defines the continuous and inclusive delivery of quality education from educational institutions over a period of time, that is able to efficiently utilize resources, with the avoidance of cutting corners on the quality of education (Albahli, 2025). AI increases educational sustainability by enhancing institutional processes, efficient resource management, and natural sustainability in education through the digital transformation.

Automating administrative operations such as; admissions, enrollment and course scheduling can be done by using AI driven systems that will correct the manual workload and errors humanly possible. Furthermore, intelligent system can help institutions optimizing the classrooms allocation, facility management, and resource planning, achieve higher efficiency and sustainability for running its institutions. Furthermore, AI enabled predictive analytics tools help universities make predictions for future enrollment rates, plan the resource allocation systematically and keep the educational quality up while avoiding wastage of resources.

In addition, AI promotes environmental sustainability by decreasing the physical use of materials caused by digitization of textbooks, resources, and assessments while contributing to green education. Together, AI driven analytics, and digitized education resources forms a sustainable and scalable model that one can use for delivery of global education.

6.4 Integration and Future Outlook

The integration of these AI-driven trends necessitates proactive policy frameworks and collaborative stakeholder engagement (Mehmood, 2022). The alignment of technological advancement with pedagogical objectives and standards of ethical consideration towards consistent equitable access for use is the responsibility of institutions. At the same time, there is an ongoing research and iterative development that is necessary to tune the AI capabilities to fit the emerging educational requirements.

The focus of the future directions should be on developing comprehensive training programs for educators and administrators who will then be able to effectively use AI technologies. Moreover, data privacy, algorithmic transparency and accessibility standard policies must also be addressed by institutional policies as a way of abating ethical concerns, or otherwise potential biases. By employing this integrated approach, AI innovations always remain consistent with the whole picture of educational objectives, inclusiveness, sustainability and global education improvement.

All in all, the advent of adaptive learning platforms, AI based virtual learning environment and sustainability conscious AI applications signify not only the technological advancement but also the change in the strategy to adopt an innovative, inclusive and resilient learning system.

7 Future Directions and Recommendations

In this regard, as Artificial Intelligence (AI) becomes firmly embedded in the education sector and specifically in higher education, stakeholders cannot reply on just embarking on one or two research projects per department or at the faculty level. All this requires strategic foresight, robust policy frameworks, and meaningful collaboration amongst stakeholders to propel sustained and inclusive educational development (Chiu, 2024). This section offers critical recommendations and insights on the future directions for the integration of AI in higher education and the necessity of elaborate policies and cooperative actions so that the possible potential of AI could be fully realized (Charles, 2020).

Strong partnerships between AI technology companies, developing AI technologies, and educational institutions are to be fostered, so that the outcome of AI is accommodative to certain pedagogical goals (Renz, 2023). Partnerships like these allow for specific AI usage in specific contexts that can then craft educational experience that can aptly serve a student's needs and at the same time, meet the objectives of the institution. Also, industry–academic collaboration is beneficial in the sense that it can offer valuable insights, practical experiences, and real-world contexts necessary in the development of such educational tools in AI.

There is the active involvement of educators in the design and development of these AI-driven educational systems in order for them to be usable, acceptable and effective (Edam, 2024). Inclusive participation does go a long way in encouraging stakeholders to speak up, offer suggestions of improvement and create innovation together, enables participants to create a sense of ownership and mutual understanding.

Collectively, the creation of multidisciplinary teams including technologists, educational experts, psychologists, ethicists, and sociologists could provide comprehensive understanding of such complexity related to AI in education. By taking multidisciplary approaches these allow technologically innovative, educationally meaningful and ethically responsible solutions to this problem.

They can also contribute to international cooperation among education institutions, government bodies and other international organizations on behalf of AI guided education initiatives globally. Cooperation with such AI is based on the ability to share knowledge, to exchange technologies, and resource pooling, and contributes to close the gap of AI adoption in development regions and widespread educational disparities.

Future joint efforts are therefore urged to promote capacity building initiatives for instance through training programmes and workshop, which will enable educators and institutional leaders to acquire essential AI literacy and competencies. Institutional empowerment can be provided to educators through skill development in utilizing AI, through which institutions are better equipped with using AI technologies, thus enriching educational experiences and outcomes.

In the field of research and innovation, collaboration is equally encouraged where stakeholders are invited to conduct joint research projects that investigate new utilizations, the resolution of pressing difficulties, and the discovery of emerging chances in AI enhanced education. Stakeholders are encouraged to wholy investing in research and development in AI education technology, allowing dedicated funding and resource to be allocated for innovations to be continued in this field.

Finally, broad and systematic synergies among developers of AI, educational institutions, policy makers, students, and industry stakeholders will help shape AI that is ethically, pedagogically and socially constructive. By default, a higher education institution should, with patience and care, collaborate with a mix of humans and machines to fully capitalise on the transformative capabilities of AI to achieve inclusive and sustainable educational futures.

8 Conclusion

The incorporation and progress of Artificial Intelligence (AI) in higher education are responsible for bringing significant transformation in educational landscape by changing traditional pedagogy models and reshaping the pathway of learning. In the last few decades, AI has emerged from a supplementary tool only to become an integral part of the structural and functional framework of current knowledge systems. In terms of educational efficacy, AI has brought around intelligent tutoring systems, adaptive assessment techniques, streamlined administrative processes and through these means it has furthered the objectives of global education including the Sustainable Development Goal 4 (SDG 4) that focusses on quality, equity and inclusion in education.

However, the contribution of AI in the field of education reaches far broader than the improvement of academic performances. It actively keeps an inclusive educational environment where learners of different nationality and different ability take tailored educational experience. Consequently, AI driven personalized learning systems have changed the classrooms to the extent that students can now learn at their own pace and at their level of preferred skill while being able to mitigate the challenges associated with the conventional 'one size fits all' educational approaches. The systems, which are vital in bringing up students, who could have been marginalized and under served by conventional education are crucial in enhancing inclusivity and equitable access to high quality education in the whole world.

In addition, AI powered intelligent tutoring systems have transformed the manner of interaction amongst the students and teachers and have assisted the learners with instant feedback and personalised support in one to one engagement that closely mimics the ability to pay attention and respond as in the cases of human teacher. Historically, the inability to afford individual attention and the existence of small class sizes has meant that this level of individual attention would not be possible. Educational institutions too have become efficient and consistent through means of automated assessment and grading systems thus cutting down on instructor time spent per student as well as time dedicated to curricular innovation and student engagement.

Although these are indeed enormous achievements, the integration of AI in the educational domain is happening quickly, and consequently, many ethical complications as well as practical challenges have arisen. These notably quite include algorithmic bias, the safeguarding of data privacy and security, that raise certain risks relative to achieving equitable education outcomes. Biases may be embedded in AI systems due to datasets that are stratified along various inequities, or from the deficiencies of the algorithms themselves with adverse effect on some student groups. The same is true for AI systems, which themselves require enormous amounts of student data to have been collected and processed. Therefore, it is the ethical and operational responsibility (also a challenge) of educational institutions, policy makers and technology providers to proactively tackle these ethical and operational challenges ahead of time, so as to maximize AI's benefits in education and minimize potential harms associated with it.

Furthermore, there is an ongoing challenge for the digital divide: the persistent divide between those that have an easy life with modern technology and the internet and those who don't. While it is promising, educational technologies that are AI driven depend significantly on extremely robust technological infrastructures as well as equal availability of internet which is not always the case. As a result, the likelihood of AI spread in education leading to more disparities instead of bridging them is quite high unless intentional, targeted actions are taken accordingly. Consequently, the equitable technology distribution, effective policy frameworks and promoting international collaboration to providing inclusive technological accessibility and affordability are, therefore, the critical necessities.

And some emerging trends in AI as well reflect the future potential of AI in the future educational landscapes as we've discussed in this talk.. AI as categorised in the learning platforms, virtual learning environment powered by AI and Sustainability AI applications suggest a transforming educational ecosystem ready to take on great changes. These innovations have started to create these resilient educational practices that are able to change on the fly to meet the diverse needs of the learner and respond to emerging societal demands in ways that will prepare learners to enter a professional landscape where knowledge and skills will be continuously being revamped very quickly.

With action from many elements—educators, policymakers, technologists, in partnership with learners—there are deliberative steps that can be taken to maximize the positive impact of AI in education looking forward. Institutional framework requirements that need to be continued to be developed and refined are strong, clear policy guidelines, and effective regulatory mechanisms. Developing transparent, public, and ethically principled AI applications requires all players in the educational arena to come together to guarantee that technology improves and not erodes the integrity or version of educational equity.

All in all, AI has limitless possibilities to evolve education from pre primary to high education to become an environment that will promote lifelong learning instead of a situation of option. Consequently, this transformative capability creates for AI not only the ability to be an important technological asset but also an essential element in attaining a truly holistic, inclusive and sustainable approach underneath ensuring a globally comprehensive education development. If fit continues on this path, AI might strongly improve the quality, affordability, and sustainability of education as more and more educational institutions adopt and refine AI applications to satisfactorily serve their stakeholders. These developments will have a great impact on global initiatives to fulfil SDG 4 and reaffirm education as an emancipating tool for human improvement, economic growth, and sustainable development.

Lastly, the way of deploying AI in higher education brings about promising and complex paths simultaneously providing opportunities and challenges. Such thoughtful and conscientious development, deployment and governance of AI educational technologies hold the promise of assisting in transforming education globally in a significant way leading to quality and equitable education for all the learners. With responsible AI, the global education community can embrace AI to transform the education system into a more adaptive, and future ready system, which will be inclusive of having dynamic challenges of the twenty-first century and beyond.

References

Adepoju, S. (2025). Transforming administrative functions through AI: Strategic planning, task automation, and resource optimization. *International Journal of Research Publication and Reviews, 6*(6), 4300–4316. https://doi.org/10.55248/gengpi.6.0125.0622

Aggarwal, D. (2023). Exploring the scope of artificial intelligence (AI) for lifelong education through Personalised & Adaptive Learning. *Journal of Artificial Intelligence, Machine Learning and Neural Network, 41*, 21–26. https://doi.org/10.55529/jaimlnn.41.21.26

Albahli, S. (2025). Advancing sustainable educational practices through AI-driven prediction of academic outcomes. *Sustainability, 17*(3), 1087. https://doi.org/10.3390/su17031087

Bardhan, R., & Kandarp, A. (2025). *Reimagining Higher Education with AI*,https://doi.org/10.21276/978-81-977736-6-2.

Bhatnagar, R. (2024). Using AI to create more equitable and inclusive higher education learning systems. *Communications on Applied Nonlinear Analysis, 32*(3s), 25–44. https://doi.org/10.52783/cana.v32.2531

Buşu, A. F. (2024). AI-powered classrooms: A revolution in learning environments. *Analele Universităţii Din Craiova Seria Ştiinţe Filologice Limbi Străine Aplicate, 2024*(1), 96–103. https://doi.org/10.52744/aucsflsa.2024.01.11

Butson, R., & Spronken-Smith R. (2024). AI and its implications for research in higher education: A critical dialogue, Higher Education Research & Development, vol. 43, no. 3, pp. 563–577, https://doi.org/10.1080/07294360.2023.2280200.

Chaney, H., Abou Najm, M., & Jose Lopez Serrano, M. (2022). Comparing academia's perception of needed SDG research to SDG Progress reports and known SDG synergies and Tradeoffs, Copernicus GmbH. Retrieved April 03, 2025, from https://doi.org/10.5194/egusphere-egu22-10867.

Charles. (2020). Research-practice partnerships in higher education: Developing tools to mediate collaboration between stakeholders, In *Proceedings of the 2020 AERA Annual Meeting*, Washington DC: AERA. Retrieved April 03, 2025, fromhttps://doi.org/10.3102/1586509

Chiu, T. K. F. (2024). Future research recommendations for transforming higher education with generative AI. *Computers and Education: Artificial Intelligence, 6*, 100197. https://doi.org/10.1016/j.caeai.2023.100197

Chouzouris, N. (2024). AI for higher education: Alternative ways of learning and risks, In *The evolution of artificial intelligence in higher education*, Emerald Publishing Limited, pp. 195–201. Retrieved April 03, 2025, from https://doi.org/10.1108/978-1-83549-486-820241012.

Conklin, T. A. (2016). Knewton (an adaptive learning platform available at https://www.knewton.com/). *Academy of Management Learning & Education, 15*(3), 635–639. https://doi.org/10.5465/amle.2016.0206

Edam, S. M. I. (2024). AI in educational design and technological development. In *Advances in Educational Technologies and Instructional Design*, IGI Global, pp. 25–60. Retrieved April 03, 2025, from https://doi.org/10.4018/979-8-3693-2728-9.ch002.

Fabiano, N., Gupta, A., Bhambra, N., Luu, B., Wong, S., Maaz, M., Fiedorowicz, J. G., Smith, A. L., & Solmi, M. (2024). How to optimize the systematic review process using AI tools. *ACAMH Learn.* https://doi.org/10.13056/acamh.31451

Ferster, B. (2022). Intelligent Tutoring Systems. In *Intelligent Tutoring Systems*, Routledge. Retrieved April 03, 2025, from https://doi.org/10.4324/9781138609877-ree6-1.

Gao, W. et al., (2023). Problem-solving using logic and reasoning, mathematics, algorithms, python and generative AI. In *2023 international conference on computational science and computational intelligence (CSCI)*, IEEE, pp. 371–377. Retrieved April 03, 2025, from https://doi.org/10.1109/csci62032.2023.00066.

Hales, R. & Birdthistle, N. (2022). The sustainable development goals—SDG#4 quality education. In *Attaining the 2030 Sustainable Development Goal of Quality Education*, Emerald Publishing Limited, pp. 1–8. Retrieved April 03, 2025, from https://doi.org/10.1108/978-1-80382-475-820221001.

Hamid, O. H. (2024). Impact of artificial intelligence (AI) in addressing students at-risk challenges in higher education (HE). In *The evolution of artificial intelligence in higher education*, Emerald Publishing Limited, pp. 183–194. Retrieved April 03, 2025, from https://doi.org/10.1108/978-1-83549-486-820241011.

Irshad, S. (2024). Role of human and artificial intelligence biases on organizational performance, efficiency, and enhanced productivity. In *Human perspectives of industry 4.0 organizations*, CRC Press, pp. 111–130. Retrieved April 03, 2025, from https://doi.org/10.1201/978103261 6810-6.

Jiao. (2024). AI-driven personalization in higher education: Enhancing learning outcomes through adaptive technologies. *Adult and Higher Education, 6*(6). https://doi.org/10.23977/aduhe.2024. 060607

Kamenskih. (2022). The analysis of security and privacy risks in smart education environments. *Journal of Smart Cities and Society, 1*(1), 17–29. https://doi.org/10.3233/scs-210114

Kinshuk. (2016). Implementation process of adaptive and personalized learning environments. In *Designing Adaptive and Personalized Learning Environments*, Routledge, 2016. | Series: Interdisciplinary: Routledge, pp. 119–132. Retrieved April 03, 2025, from https://doi.org/10. 4324/9781315795492-15.

Kurtzke, S. (2024). Integrating AI literacy into higher education. In *Artificial intelligence applications in higher education*, Routledge, pp. 275–301. Retrieved April 03, 2025, from https://doi. org/10.4324/9781003440178-16.

Lata, P. (2024). Towards equitable learning: Exploring artificial intelligence in inclusive education. *SSRN Electronic Journal*. https://doi.org/10.2139/ssrn.4987156

Mariotti, C. (2019). The role of the UK in tackling global inequality: Contributions to achieving SDG 10, Oxfam GB. Retrieved April 03, 2025, from https://doi.org/10.21201/2019.4467.

Mehmood, J., Hajdini, L., Iaia, F. D. L., & Sakka, G. (2022). Stakeholder engagement and SDGs: The role of social media in the European context. *EuroMed Journal of Business, 18*(1), 111–128. https://doi.org/10.1108/emjb-11-2021-0173

Kristina Posavec. (2025). AI tools and platforms for personalized learning, In *Advances in Educational Technologies and Instructional Design*, IGI Global, pp. 25–60. Retrieved April 03, 2025, from https://doi.org/10.4018/979-8-3693-3474-4.ch003.

Potluri, R. M. & Kilaru, M. (2024). Emerging trends of AI Technologies in the Higher Education Sector, In A*dvances in Educational Technologies and Instructional Design*, IGI Global, pp. 41–56. Retrieved April 03, 2025, from https://doi.org/10.4018/979-8-3693-7220-3.ch003.

Pulivarthy, P. & Whig, P. (2024). Bias and fairness addressing discrimination in AI systems, In *Advances in Human and Social Aspects of Technology*, IGI Global, pp. 103–126. Retrieved April 03, 2025, from https://doi.org/10.4018/979-8-3693-4147-6.ch005.

Renz. (2023). AI in education: Educational technology and AI. In *Work and AI 2030*, Springer Fachmedien Wiesbaden, pp. 353–360. Retrieved April 03, 2025, from https://doi.org/10.1007/ 978-3-658-40232-7_39.

Singh, S. & Singh, N. (2023). GPT-3.5 vs. GPT-4, unveiling OpenAI's latest breakthrough in language models, Institute of Electrical and Electronics Engineers (IEEE). Retrieved April 03, 2025, from https://doi.org/10.36227/techrxiv.24486214

Sparrow, M. J. A bridge to a smart start, Northeastern University Library. Retrieved April 03, 2025, from https://doi.org/10.17760/d20247343.

Tariq, M. U. (2024). Exploration of AI precision in higher education. In *Advances in Higher Education and Professional Development*, IGI Global, pp. 273–298. Retrieved April 03, 2025, from https://doi.org/10.4018/979-8-3693-4074-5.ch011.

Taylor, J. Digital tools, rhetoric, and meaning-making: A critical exploration of addressing the digital divide and accessibility to improve our digital landscape, Iowa State University. Retrieved April 03, 2025, from https://doi.org/10.31274/td-20240329-241.

Temper, M., Tjoa, S., & David, L. (2025). Higher education act for AI (HEAT-AI): A framework to regulate the usage of AI in higher education institutions. *Frontiers in Education, 10*. https:// doi.org/10.3389/feduc.2025.1505370

Tobler, S. (2023). Smart grading: A generative ai-based tool for knowledge-grounded answer evaluation in educational assessments, Elsevier BV. Retrieved April 03, 2025, from https://doi.org/10.2139/ssrn.4637633.
Yadav, S. (2024). Artificial intelligence (AI) integration in higher education, In *Impact of artificial intelligence on society*, Chapman and Hall/CRC, pp. 43–59. Retrieved April 03, 2025, from https://doi.org/10.1201/9781032644509-4.
Yadav, M., Singh, S. K., Sharma, P., & Hung, T. H. (2024). AI and lifelong learning in developing countries," In *Advances in Educational Technologies and Instructional Design*, IGI Global, pp. 49–72. Retrieved April 03, 2025, from https://doi.org/10.4018/979-8-3693-0884-4.ch003.
Zhou, X., & ChemistryViews. (2023). AI helps spend more time creating good ideas instead of communicating them. *ChemViews*. https://doi.org/10.1002/chemv.202300107

The Importance of Evidence-Based Policy Guidelines in the Age of AI, Educational Technologies, and Agile Methodologies

José Luiz Esteves

Abstract The rapid evolution of artificial intelligence (AI), educational technologies, and agile methodologies necessitates a paradigm shift in pedagogical approaches to accommodate multigenerational learning environments. This article advocates evidence-based educational policies that integrate scientific research, technological advancements, and agile frameworks to enhance learning outcomes across diverse age groups. We propose four key pillars for policy development: (1) the establishment of an **Observatory of Best Practices**, (2) the promotion of applied research collaborations, (3) the integration of **agile governance models**, and (4) a **structured framework for educational policy formulation and legislative review**. Additionally, we examine **evaluation and monitoring frameworks** for educational programs and discuss **institutional strengthening through participatory governance**. The article incorporates recent research by the author on **AI applications in special education** and **Freirean pedagogy in digital learning**, demonstrating how these approaches can inform inclusive policymaking. Case studies, including hybrid learning models (OECD) and Brazil's SAEB-driven literacy policies, illustrate the effectiveness of data-driven decision-making.

Keywords AI in education · Agile methodologies · Educational policy · Evidence-based education · Inclusive education · Program evaluation · Multigenerational learning

J. L. Esteves (✉)
IEA/USP Chairs and ABES Think Tank, Rio de Janeiro, Brazil

COPPE/UFRJ, Rio de Janeiro, Brazil

A. Ara (ed.), *Role of Artificial Intelligence in Higher Education*,
https://doi.org/10.1007/978-981-95-5350-1_3

1 Introduction

The digital transformation of education requires comprehensive policy frameworks that address **multigenerational learning needs**, **technological integration**, and **legislative adaptability**. This article expands previous discussions by incorporating:

1. **A Model for Formulating and Revising Educational Legislation**
2. **Recent research on AI applications in special education** (Esteves, 2025)
3. **Freirean pedagogy in digital learning contexts** (Esteves, 2023)

These additions provide a more holistic view of contemporary educational challenges and solutions.

1.1 Current Summary of Studies and Research on Instruments, Parameters and Processes for Evaluation and Monitoring of the Management of Priority Educational Programs

The evaluation and monitoring of educational programs are not merely bureaucratic exercises—they are essential mechanisms for ensuring that public investments in education translate into meaningful outcomes. As educational systems face increasing pressure to deliver equitable, high-quality learning opportunities, the ability to assess the effectiveness, efficiency, and fairness of programs becomes a cornerstone of sound policy and governance.

Recent research underscores the importance of combining **quantitative and qualitative methodologies** to capture the full complexity of educational interventions. Three key approaches have emerged as particularly influential: **education quality indicators**, **monitoring tools**, and **evidence-based evaluation methodologies**.

- **Education Quality Indicators**: Education quality indicators serve as benchmarks for assessing the performance and inclusiveness of educational systems. These indicators, developed by international organizations such as UNESCO and the OECD, typically include metrics such as enrollment and completion rates, student performance on standardized assessments, teacher qualifications, and equity in access to learning opportunities.
  (a) **Brazilian Example:** The *IDEB* (Índice de Desenvolvimento da Educação Básica) is a composite index used in Brazil to measure the quality of basic education. It combines student performance on standardized tests with school progression rates, offering a snapshot of both learning outcomes and system efficiency. Policymakers use IDEB scores to identify underperforming schools and allocate targeted support.
  (b) **International Example:** The OECD's *Education at a Glance* report provides comparative data on education systems across member countries. It includes

indicators on student achievement, funding, teacher workload, and access to early childhood education—enabling countries to benchmark their progress and identify areas for reform.

- **Monitoring Tools**: Monitoring tools are essential for the continuous collection, analysis, and use of educational data. Among the most widely adopted is the **Education Management Information System (EMIS)**, which supports data-driven decision-making at all levels of the education system.

  Functionality: EMIS platforms typically track student enrollment, attendance, teacher deployment, infrastructure, and financial flows. When properly implemented, they allow for real-time monitoring of program implementation and help identify bottlenecks or disparities.

  Brazilian Context: The *Sistema Integrado de Monitoramento, Execução e Controle (SIMEC)* is a digital platform used by Brazil's Ministry of Education to manage and monitor federal education programs. It integrates data from various sources, enabling more transparent and efficient oversight of initiatives such as school construction, textbook distribution, and teacher training.

  International Example: In **Uganda**, the Ministry of Education uses EMIS to monitor school-level data and inform national planning. The system has been instrumental in identifying regions with low enrollment rates and guiding the allocation of resources to underserved communities
- **Evidence-Based Methodologies**: To move beyond descriptive statistics and understand what truly works, researchers and policymakers are increasingly turning to **evidence-based evaluation methods**. Among the most rigorous are **Randomized Control Trials (RCTs)** and **Impact Evaluations**, which aim to isolate the causal effects of specific interventions.

  These methodologies help determine whether a program is effective, for whom, and under what conditions. They also provide insights into cost-effectiveness, scalability, and unintended consequences.

  (a) **Brazilian Example:** The *Educação Conectada* program (Connected Education Program), which promotes digital inclusion in public schools, has been the subject of impact evaluations conducted in partnership with research institutions. These studies assess not only improvements in digital literacy but also broader effects on student engagement and academic performance.

  (b) **International Example:** The *Teaching at the Right Level (TaRL)* initiative, originally developed in India, has been rigorously evaluated through RCTs. The findings have shown significant gains in foundational literacy and numeracy, leading to its adoption in several African countries and influencing global education policy debates.

2 The Need for Evidence-Based Educational Policies

2.1 Conceptual Framework: Educational Governance and Institutional Strengthening

Effective education systems are built on strong governance. This encompasses more than just managing school, it involves shaping a culture of transparency, responsiveness, and accountability. Good governance ensures that decisions aren't made in a vacuum but rooted in the realities of those directly impacted (See Fig. 1).

Three intertwined aspects of **decentralization and school autonomy**, **social participation**, and **results-oriented management** will serve as critical levers for transforming educational outcomes. Educational governance is one of the fundamental pillars for effective management of educational systems. Technical reports and strategic documents suggest that good governance involves:

- **Decentralization and School Autonomy**: Decentralized educational models allow for greater autonomy of schools and networks, resulting in policies better adapted to local realities. When educational decisions are made closer to those they affect, outcomes tend to improve. Decentralization empowers local leaders, educators, and communities to make contextually relevant choices. School autonomy allows institutions to craft strategies that reflect the needs, strengths, and aspirations of their students.

    (a) **Brazilian Perspective:** One practical example is the *Programa Dinheiro Direto na Escola (PDDE).* Rather than routing all funds through layers of bureaucracy, PDDE delivers resources straight to schools, allowing them to address urgent needs—from basic repairs to acquiring essential materials. It promotes not only efficiency but also trust in grassroots decision-making.

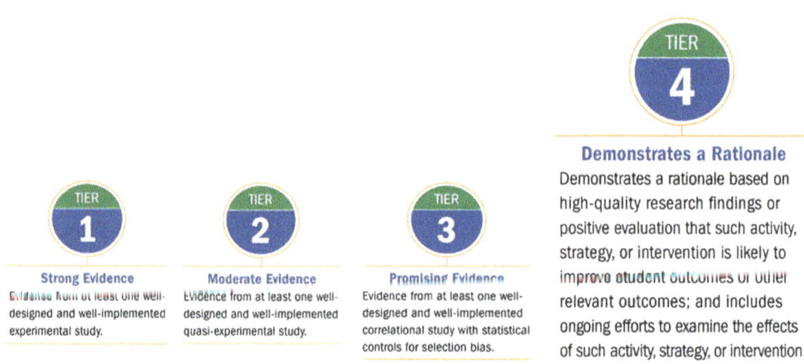

Fig. 1 EBP's four tiers of evidence. (*Source* Leveraging Evidence-Based Practices for School Improvement, Non-Regulatory Guidance: Using Evidence to Strengthen Education Investments. U.S. Department of Education, September 28, 2023)

(b) **International Insight:** In Finland, educational autonomy isn't just a policy—it's a philosophy. Schools and municipalities shape their own curricula, and teachers are given professional freedom to innovate in the classroom. This trust-based model fosters creativity and has been a key factor in the country's educational success on global platforms like the PISA assessments.

- **Social Participation**: The implementation of school councils and participatory bodies strengthens democratic management and transparency in public policies. Sustainable education reform depends on more than policies—it hinges on voices being heard. Active community engagement strengthens democratic values and aligns school practices with societal expectations.

  (a) **In Brazil:** *Conselhos Escolares* (School Councils) exemplify participatory governance. Composed of parents, teachers, students, and other community members, these councils hold real influences, approving school plans, supervising budgets, and guiding pedagogical priorities. Their presence encourages transparency and builds a collective sense of ownership.
  (b) **Globally:** In South Africa, *School Governing Bodies (SGBs)* have a statutory role, ensuring that parents, teachers, and learners shape decisions on everything from hiring staff to setting school rules. This inclusive approach helps align educational institutions with community values and needs.

- **Results-Based Management**: Tools such as **Monitoring and Evaluation (M&E)** of educational outcomes are crucial for data-informed decision-making. Governing results means using concrete evidence, such as data, evaluations, and feedback as to drive policy decisions and instructional strategies. It turns education management into a dynamic process of planning, testing, and improving.

  (a) **Brazilian Application:** Through the *Sistema de Avaliação da Educação Básica (SAEB)*, data on student performance is collected at scale. The results inform policies at various levels, from federal planning to school-level interventions. Programs like *Mais Alfabetização* use these insights to provide targeted support where learning gaps are most pronounced.
  (b) **International Practice:** The UK-based *Education Endowment Foundation (EEF)* is a leading example. Its rigorous studies identify what really works in the classroom, turning findings into accessible guidance for educators. Teachers can consult its toolkit to implement strategies backed by solid research, moving from intuition to impact.

Improving governance isn't just about better administration, it's about creating systems that listen, adapt, and grow. Whether through the autonomy of a rural school, the voice of a parent in a local council, or a policymaker drawing insights from performance data, these elements together help reshape education into something more equitable and effective.

3 Developing Guidelines and Recommendations

3.1 Proposal to Support the Development of Guidelines and Recommendations for Evidence-Based Educational Policies

Designing effective educational policies requires more than good intentions: it demands a foundation built on rigorous evidence and proven practices. To ensure that policy decisions are both impactful and sustainable, it is essential to develop clear, actionable guidelines rooted in research and real-world experience. This process involves not only identifying what works but also creating mechanisms to share, adapt, and implement these insights across diverse educational contexts(see Fig. 2).

Three (3) strategic proposals can support this endeavor: the establishment of best practices observatory, the promotion of applied research, and the use of practical, context-sensitive examples.

- **Creation of a Best Practices Observatory**: A repository of effective educational policies can help managers and policymakers. A Best Practices Observatory would serve as a centralized, dynamic repository of successful educational policies and interventions. Its purpose is to collect, curate, and disseminate evidence-based strategies that have demonstrated positive outcomes in various educational settings.

  Such an observatory could include case studies, policy briefs, implementation guides, and evaluation reports. It would be particularly valuable for education managers and policymakers seeking to replicate or adapt effective models to their own contexts.

  (a) **Brazilian Example:** The *Observatório da Educação* (Education Observatory), supported by CAPES and INEP, has played a key role in monitoring and analyzing educational indicators in Brazil. By providing open access to data and research findings, it supports more informed decision-making at municipal, state, and federal levels.
  (b) **International Example:** The OECD's *Education Policy Outlook* offers comparative insights into education reforms across member countries. It highlights policy trends, identifies successful strategies, and provides tools for countries to assess and improve their own systems.

- **Incentives for Applied Research**: Partnerships between universities, research centers, and governments can strengthen evidence-based decision-making. Bridging the gap between academic research and policy implementation is a persistent challenge. Encouraging applied research, the research that directly informs practice, can help ensure that educational policies are grounded in real-world needs and conditions.

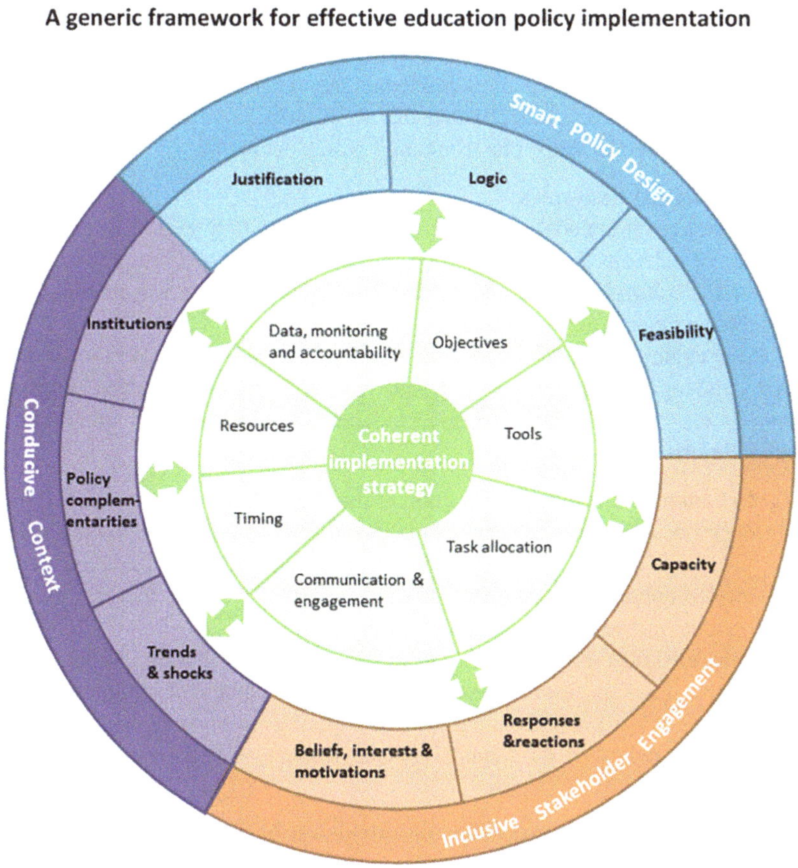

Fig. 2 A generic framework for effective education policy implementation. (*Source* OECD Education Working Paper, no. 162. OECD Publishing, Paris.)

This requires fostering partnerships between universities, research institutions, and government agencies. Funding mechanisms, research grants, and collaborative platforms can incentivize scholars to focus on pressing educational issues and produce findings that are both rigorous and actionable.

(a) **Brazilian Example:** The *Programa de Apoio à Pós-Graduação (PROAP)* and *Programa de Apoio à Pós-Graduação Stricto Sensu (PROEX)*, both coordinated by CAPES, support graduate research in education. These programs have enabled the development of studies that inform public policy, particularly in areas such as literacy, teacher training, and curriculum development.

(b) **International Example:** In Canada, the *Knowledge Network for Applied Education Research (KNAER)* brings together researchers, educators, and policymakers to co-create knowledge and translate research into practice. This collaborative model ensures that research findings are not only published but also used to shape classroom and policy decisions.

- **Other Practical Examples**:

  Abstract recommendations often fall short when they lack relevance to the realities of schools and communities. Including practical examples, grounded in specific cultural, social, and institutional contexts, makes guidelines more relatable and easier to implement.

  These examples can illustrate how policies have been adapted to local needs, what challenges were encountered, and what lessons were learned. They also serve as a source of inspiration and guidance for educators and administrators navigating similar issues.

(a) **Brazilian Example:** The Programa Ensino Médio Inovador (ProEMI) introduced flexible curricular models in Brazilian high schools, allowing schools to tailor learning paths to student interests and regional characteristics. Its implementation across different states revealed valuable insights into how national policies can be adapted locally.

(b) **International Example:** In Kenya, the Tusome Early Grade Reading Program has successfully improved literacy outcomes by combining teacher training, instructional materials, and ongoing monitoring. Its success has been attributed to its adaptability and the use of real-time data to refine implementation strategies.

Developing robust guidelines for educational policy is not a one-size-fits-all endeavor. It requires a thoughtful blend of evidence, collaboration, and contextual sensitivity. By investing in observatories of best practices, incentivizing applied research, and grounding recommendations in practical examples, education systems can move toward more effective, equitable, and sustainable reforms.

4 Evaluation and Monitoring of Educational Programs

The evaluation of programs in the educational sector should be seen not merely as a control mechanism but as a continuous process of institutional learning. In line with the results-based management approach, it is crucial that evaluation systems incorporate qualitative and contextual dimensions, respecting the territorial and cultural specificities of educational networks (see Fig. 3).

According to Stufflebeam and Shinkfield (2007), educational evaluation should be guided by criteria of utility, feasibility, accuracy, and ethics. This means that the indicators used should reflect not only academic performance but also aspects such as

Fig. 3 Drivers of Successful Data Collection and Use. (*Source* Digital Tools for Real-Time Data Collection in Education. By Emily Gustafsson-Wright, Sarah Osborne, and Muskan Aggarwal, Brookings.edu, Nov. 2022)

equity, inclusion, and student well-being. The integration of mixed methodologies—such as case studies, focus groups, and longitudinal data analysis—allows for a richer and more contextualized understanding of the impacts of public policies.

Furthermore, the adoption of digital monitoring systems, such as EMIS, can enhance the real-time collection and analysis of data, facilitating evidence-based decision-making. However, it is essential to ensure the technical training of managers and transparency in the use of this information, avoiding the political instrumentalization of data.

5 Educational Governance and Institutional Strengthening

Contemporary educational governance requires a model that combines decentralization, social participation, and institutional innovation, as pointed out and confirmed prior by UNESCO (2021) and later by OECD (2022). The Brazilian experience with school councils and education forums demonstrates that democratic management is a viable and necessary path for strengthening public educational policies.

According to Ball (2012), educational governance should be understood as a field of symbolic and material disputes, where different actors—state, civil society, private sectors negotiate meanings and priorities. In this context, institutional strengthening involves valuing active listening, transparency in decision-making processes, and horizontal accountability.

The creation of observatories of best practices, as proposed by Esteves (2025), can act as catalysts for innovation and interinstitutional learning. These spaces should be conceived as collaborative networks, involving universities, public managers, teachers, and students, with the aim of systematizing, validating, and disseminating successful experiences.

6 A Framework for Formulating and Revising Educational Legislation

In an era marked by rapid technological evolution and shifting societal expectations, educational legislation must be more than a static set of rules—it must be a living framework, capable of adapting to change while remaining grounded in democratic values and scientific evidence. The formulation and revision of legal frameworks in education require a model that is at once structured, inclusive, and responsive to emerging realities.

This proposed framework is built upon four interdependent pillars: **identifying legislative gaps**, **expanding social participation**, **leveraging international benchmarking**, and **ensuring legislative agility**. Together, these elements form a dynamic approach to crafting laws that are both forward-looking and contextually grounded.

The formulation and review of legal frameworks in education require a model that is simultaneously structured, participatory and responsive to social, technological and pedagogical transformations. In times of accelerated change, such as those caused by artificial intelligence, the digitalization of education and new demands for lifelong learning, it is essential that educational legislative systems incorporate mechanisms for continuous updating, social listening and alignment with scientific evidence.

Each of these axes is detailed below:

6.1 Identifying Gaps in Current Legislation

The first step in legislative reform is recognizing where existing laws fall short. This involves a systematic review of current statutes to detect inconsistencies, outdated provisions, or areas that fail to address contemporary educational challenges—such as digital inclusion, data privacy in schools, or the integration of AI in learning environments.

A critical analysis of current education legislation is the first step to ensuring its relevance and effectiveness. In Brazil, the **Law of Guidelines and Bases for National Education (LDB)**, enacted in 1996, represented a significant advance by consolidating principles such as democratic management, the valorization of teaching, and the articulation between levels of education. However, after almost three decades, new demands have emerged — such as hybrid education, digital inclusion, the protection of educational data, and the ethical use of artificial intelligence in the classroom — that are still not fully covered by the LDB.

A practical example of this gap is the lack of clear guidelines on the use of algorithms in assessment processes and personalization of learning, a topic already regulated in countries such as **Estonia** and **Finland**. These gaps can be identified through legislative audits, comparative analyses, and technical consultations with experts in public policy and education law.

In **Germany**, the federal structure of education has led to disparities in digital education policies across states. A national review in 2020 identified gaps in digital infrastructure and teacher training, leading to the *DigitalPakt Schule*, a federal initiative to modernize schools and bridge legislative inconsistencies.

6.2 *Public Consultation and Social Dialogue*

Laws gain legitimacy and relevance when they are shaped by those they affect. Expanding social participation means creating meaningful opportunities for students, educators, families, and civil society to contribute to the legislative process. This can take the form of public consultations, participatory forums, or digital platforms for feedback.

The construction of legitimate and effective educational policies necessarily involves actively listening to the stakeholders of education. Teachers, students, administrators, families and social movements must be protagonists in the legislative review process. In **Brazil**, the experience of the 2014 **National Education Plan (PNE)**, which had broad popular participation in its preparatory conferences, is an example of how social dialogue can enrich the content and legitimacy of educational standards.

In addition, the use of digital public consultation platforms, such as the **Participa + Brasil** portal, can expand the reach and diversity of voices heard, especially in contexts of territorial inequality. Active listening must be continuous and not just occasional, allowing legislation to reflect the real transformations in society.

In **New Zealand**, the Ministry of Education's *Education Conversation | Kōrero Mātauranga* initiative invited citizens to co-design the future of education. Through town halls, surveys, and targeted outreach to underrepresented groups, the process ensured that legislative reforms reflected a wide spectrum of voices.

6.3 *International Benchmarking*

Comparing successful international experiences is a powerful strategy for inspiring legislative innovation. **Finland**, for example, revised its national curriculum in 2016 based on a cross-cutting and interactive approach, incorporating socio-emotional skills, computational thinking and phenomenon-based learning. **South Korea** has implemented a robust regulatory framework for the integration of educational technologies, with a focus on teacher training and digital equity.

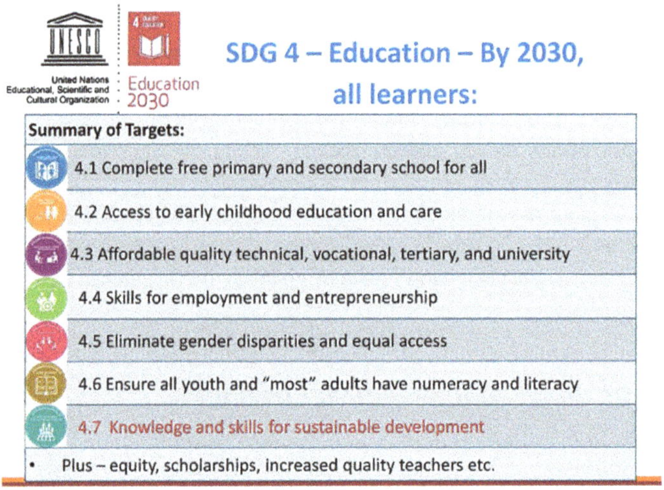

Fig. 4 UNESCO's SDG 4 Summary of Targets. (*Source* Education 2030 Framework for Action. Regional Conference Towards a Common Vision for Agenda 2030 in Central Asia Presentation, 2–4 November 2016, Almaty, Kazakhstan)

Brazil's alignment with UNESCO's Education 2030 Agenda has influenced national policies on inclusive education and early childhood care. Comparative studies with countries like **Portugal** have informed debates on curriculum reform and teacher evaluation. In the Latin American context, **Chile** has stood out for creating a Superintendence of Education that acts as an independent regulatory body, ensuring compliance with educational standards with a focus on quality and equity.

Mongolia's recent overhaul of its education law, supported by UNESCO, incorporated international benchmarks to ensure alignment with global standards on equity, quality, and lifelong learning. This process not only modernized the legal framework but also reinforced the country's commitment to education as a human right.

These examples demonstrate that benchmarking should not be a simple import of models, but a critical and contextualized adaptation to local realities (see Fig. 4).

6.4 *Legislative Agility*

In a world where educational needs evolve rapidly—driven by artificial intelligence, climate change, and labor market shifts—laws must be designed with built-in flexibility. Legislative agility refers to the capacity of legal systems to respond quickly to new developments without sacrificing due process or democratic oversight.

The rigidity of legislative processes often prevents the education system from responding quickly enough to changes in the contemporary world. Therefore, it is essential to incorporate mechanisms for **legislative agility**, such as mandatory periodic reviews, sunset clauses (automatic expiration of standards), and small-scale regulatory tests (pilots).

In **Brazil**, the experience of the "Novo Ensino Médio", with its pilot phase in state networks, also illustrates the importance of testing and adjusting policies before their large-scale implementation. The COVID-19 pandemic in Brazil also exposed the rigidity of some educational regulations, particularly around school calendars and remote learning. In response, emergency measures were enacted to allow for hybrid instruction and alternative assessment models, highlighting the need for more adaptable legal frameworks moving forward.

An inspiring example is the **"regulatory sandbox"** model adopted by the **United Kingdom** and Singapore, which allows new educational practices to be tested in controlled environments before their final regulation. **Estonia**'s education legislation for instance, includes provisions for periodic review and digital innovation. The country's agile legal approach enabled a swift transition to online learning during the pandemic and continues to support experimentation with AI-driven tools in classrooms.

Educational legislation must be more than a reflection of the present, as it must somehow anticipate the future. By identifying legal blind spots, engaging society in meaningful dialogue, learning from global peers, and embracing flexibility, countries can ideally craft laws that are resilient, inclusive, and responsive to the demands of a changing world.

7 AI and Inclusive Education: Insights from Recent Research

The integration of artificial intelligence (AI) into educational contexts has significantly expanded the possibilities for personalized learning, particularly for students with specific educational needs. The World Bank (2020) has considered during the covid 19 global pandemic that education technology should be considered a solid instrumental to learning acceleration and inclusion. However, the adoption of such technologies must be guided by ethical, pedagogical, and inclusive principles to ensure that technological advancements do not exacerbate existing inequalities.

The integration of artificial intelligence (AI) into education is reshaping how we understand and support learning, and especially for students with diverse needs. Recent studies suggest that AI holds enormous potential to personalize instruction, enhance accessibility, and reduce barriers for learners who have historically been marginalized. Yet, this promise is not without its caveats. The deployment of AI in educational settings must be guided by a framework that is not only technologically sound but also ethically grounded, pedagogically informed, and deeply inclusive.

AI is not a silver bullet at all! But when thoughtfully implemented, it can be a powerful ally in the pursuit of inclusive education. Its success depends on a delicate balance: leveraging technological innovation while honoring human relationships at the heart of learning. As research continues to evolve, so too must our commitment to ensuring that AI serves all learners, especially those who have been historically left behind.

The following section presents recent evidence on the use of AI in inclusive education, with an emphasis on approaches grounded in Freirean pedagogy and applications for Autism Spectrum Disorder (ASD).

7.1 AI Applications for Autism Spectrum Disorder (ASD)

The use of AI-based systems to support students with Autism Spectrum Disorder (ASD) has shown promising results, especially in customizing learning routines, developing socio-emotional skills, and mediating communication. Technologies such as virtual assistants, emotion recognition software, and adaptive learning platforms have been employed to create more responsive learning environments tailored to neurodiverse needs. As research and practice evolve, AI-based tools are proving to be valuable allies in addressing the unique learning profiles of neurodiverse students. These technologies are not only enhancing academic engagement but also fostering communication, emotional development, and autonomy.

To fully understand the scope of AI's impact, it is helpful to explore three key areas where its application has been most promising: **customization of learning routines**, **development of socio-emotional skills**, and **communication mediation**.

- **Customizing Learning Routines**

Students with ASD often benefit from structured, predictable environments and individualized learning pathways. AI-powered platforms can analyze behavioral patterns, learning preferences, and performance data to adapt content delivery in real time. This level of personalization helps reduce anxiety, increase focus, and promote sustained engagement.

(a) **Brazilian Example:** In Recife, a pilot project led by the Federal University of Pernambuco introduced an AI-based scheduling assistant for students with ASD in public schools. The tool allowed teachers to create visual timetables that adjusted dynamically based on student responses and classroom conditions. Early findings indicated improved attendance and task completion rates.
(b) **International Example:** In the Netherlands, the *Zeno AI Tutor* has been used to support children with ASD by offering personalized math and reading exercises. The system adjusts difficulty levels and pacing based on the learner's interaction patterns, helping to maintain motivation and reduce frustration.

- **Developing Socio-Emotional Skills**

AI technologies are also being used to support the development of emotional recognition, empathy, and social interaction, and these areas can be particularly challenging for individuals on the autism spectrum. Emotion recognition software, for instance, can help students interpret facial expressions and social cues in a safe, non-judgmental environment.

(a) **Case in Point:** The *Empathic Computing Lab* in Japan has developed wearable AI devices that provide real-time feedback on emotional states during peer interactions. These tools help students with ASD recognize when a conversation partner is confused, bored, or engaged, allowing them to adjust their behavior accordingly.
(b) **Brazilian Context:** In São Paulo, a collaboration between SENAI and local edtech startups has led to the development of gamified platforms that use AI to simulate social scenarios. Students practice responding to various emotional cues, with the system offering gentle corrections and encouragement.

- **Mediating Communication**

Communication barriers are among the most significant challenges faced by individuals with ASD. AI-driven augmentative and alternative communication (AAC) tools are helping bridge this gap by translating gestures, images, or typed input into spoken language. These tools empower students to express their needs, participate in class discussions, and build relationships.

(a) **Global Example:** Microsoft's *Seeing AI* and Google's *Look to Speak* are being adapted for use in classrooms to support non-verbal students. These tools use computer vision and eye-tracking to enable communication through gaze or object recognition, offering new levels of independence.
(b) **Brazilian Initiative:** The *Voz Ativa* project, developed by researchers at the Federal University of Minas Gerais, integrates AI with speech synthesis to assist non-verbal students in mainstream classrooms. Teachers report that the tool has improved peer interaction and reduced behavioral incidents linked to communication frustration.

Recent published research (Esteves, 2025) demonstrates that the effectiveness of these technologies increases significantly when integrated with Freirean pedagogical principles—such as respect for the learner's individuality, dialogue as the foundation of mediation, and the valorization of lived experience as a starting point for knowledge construction. In a case study conducted at a public school in São Paulo, the implementation of an adaptive AI system, combined with dialogical pedagogical practices, led to measurable improvements in autonomy and participation among students with ASD.

Moreover, international initiatives such as the European Union's "Empathic" project and the use of the NAO robot in schools in France and Japan illustrate how AI can be used to foster mediated social interactions and develop communication skills in autistic children. These experiences reinforce the importance of a student-centered approach that combines technological innovation with pedagogical sensitivity.

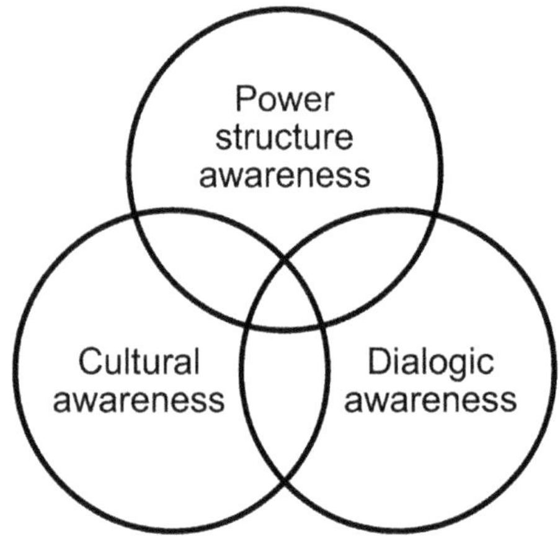

Fig. 5 The importance of Freirean Pedagogy for knowledge construction (*Source* Handbook of Communication for Development and Social Change. Springer, Singapore)

7.2 *Freirean Pedagogy in Digital Learning*

Freirean pedagogy, with its emphasis on conscientization, dialogue, and social transformation, offers a robust theoretical framework for constructing critical educational practices in digital environments. Rather than reproducing traditional instructional models mediated by technology, the Freirean approach invites the creation of virtual learning spaces that are interactive, collaborative, and politically engaged.

In online education contexts, active methodologies such as project-based learning, problematization, and digital culture circles have been successfully used to promote student agency and collective knowledge construction. As presented by Esteves (2023) at the UNIP University PD Working Group Conference Meeting, integrating digital technologies with critical pedagogy not only expands access to education but also redefines the role of the student as a historical subject and agent of transformation.

Practical examples include the use of platforms like Moodle and Google Classroom to create thematic discussion forums, collaborative production of educational podcasts and videos by students, and the use of digital mapping tools for critical analysis of local territories and realities. When guided by a Freirean perspective (see Fig. 5), these practices contribute to overcoming the "banking model" of education and to building an emancipatory digital culture.

8 Case Studies

The incorporation of case studies in the design and evaluation of educational policies allows for the identification of critical success factors and the adaptation of solutions to different contexts.

- In Brazil, the experience of the **Ceará State Public Schools System** with evidence-based literacy policies is an emblematic example. The systematic use of SAEB data, combined with continuous teacher training and results-oriented school management, has led to significant improvements in learning indicators (Cruz et al., 2021).
- Internationally, the **Finnish** model of curricular decentralization and teacher valorization continues to be a reference. South Korea stands out for its strategic integration of educational technologies, with robust investments in digital infrastructure and teacher training for the pedagogical use of AI.

These cases show that there are no single solutions, but rather guiding principles—such as equity, participation, and innovation—that can be adapted to different realities. The systematization of these experiences, through open repositories and collaborative platforms, is essential for building more responsive and effective public policies.

9 Conclusion and Policy Recommendations

Faced with the challenges posed by the digital age, climate emergency, and structural inequalities, the formulation of evidence-based educational policies requires more than technique: it requires institutional courage. Courage to break with obsolete models, to listen to the subjects of education, and to experiment with new ways of teaching and learning.

The recommendations presented in this article converge towards the construction of a more just, innovative, and sustainable educational ecosystem. Among them, the following stand out:

1. Develop flexible legislative frameworks, with mechanisms for periodic review and expanded social participation
2. Invest in inclusive technologies, focusing on personalized learning and respect for diversity
3. Apply Freirean pedagogical principles in digital education, promoting dialogue, awareness, and social transformation
4. Strengthen the culture of formative evaluation, focusing on continuous improvement and institutional learning
5. Create interinstitutional collaboration networks, for the sharing of best practices and innovation in public policies

The integration of AI into educational strategies for students with ASD is not merely a technological advancement. It is a step towards a more humane, inclusive, and responsive learning environment. However, the success of these tools depends on thoughtful implementation, ethical safeguards, and ongoing collaboration between educators, families, technologists, and the students themselves.

As we continue to explore the potential of AI in education, it is essential to ensure that innovation serves equity, and that every learner, regardless of what any neurological profile could thrive.

As Paulo Freire (1996) reminds us, "Education does not transform the world. Education changes people. People transform the world."

May the educational policies of the twenty-first century rise to this challenge.

References

Ball, S. J. (2012). *Global education Inc: New policy networks and the neo-Liberal imaginary*. Routledge.

Cruz, R. S., Oliveira, A. M., & Pinto, J. M. (2021). *Políticas de alfabetização no Ceará: evidências e resultados*. Revista Brasileira de Educação, 26.

Esteves, J. L. (2025). *AI applications in special education: A Freirean approach*. Schreiben Editora.

Esteves, J. L. (2023). *Paulo Freire's pedagogy and digital technologies seminar presentation. UNIP PD working group conference proceedings*. Brazil.

Freire, P. (1996). *Pedagogia da Autonomia: Saberes Necessários à Prática Educativa*. Paz e Terra.

OECD. (2022). *Education policy outlook 2022: Transforming pathways for lifelong learners*. OECD Publishing.

Stufflebeam, D. L., & Shinkfield, A. J. (2007). *Evaluation theory, models, and applications*. Jossey-Bass.

UNESCO. (2021). *Reimagining our futures together: A new social contract for education*. UNESCO.

World Bank. (2020). *The Promise of Education Technology to Accelerate Learning for All*. Washington, DC.

Investigating Possibilities for Sustainability in Higher Education by Using Cutting-Edge Digital Technologies: Implications for Emerging Economies

Kamalakanta Muduli, Ravindra Luhach, Raj Kumar, Bikash Ranjan Moharana, and Adimuthu Ramassamy

Abstract Sustainability in Higher Education is a worldwide issue that falls under "Sustainable Development Goal: 4" (SDG#4), aims to provide high-quality education. It is one of the seventeen universal SDGs established by the United Nations. The article explored the principles of sustainable development, emphasizing the significance of incorporating sustainability into higher education. Additionally, the research examined several instances of successful outcomes and provided guidance on strategizing the incorporation of sustainability practices in higher educational institutions. The article further investigated the utilization of sophisticated digital technologies viz. AI, IoT, Big Data, Cloud Computing etc. in higher education, including possibilities, difficulties it presents, as well as recommendations for improving procedures and their consequences for developing nations like Papua New Guinea (PNG).

Keywords Sustainable development goal (SDG) · Higher education · Artificial intelligence (AI) · Internet of things (IoT)

K. Muduli (✉) · B. R. Moharana
School of Mechanical Engineering, Papua New Guinea University of Technology, Lae, Papua New Guinea
e-mail: kamalakanta.muduli@pnguot.ac.pg

R. Luhach · R. Kumar
School of Electrical and Communication Engineering, Papua New Guinea University of Technology, Lae, Papua New Guinea

A. Ramassamy
School of Business Studies, Papua New Guinea University of Technology, Lae, Papua New Guinea

A. Ara (ed.), *Role of Artificial Intelligence in Higher Education*,
https://doi.org/10.1007/978-981-95-5350-1_4

1 Introduction

The graduates who possess a comprehensive understanding of sustainability can create projects that effectively reduce negative consequences, take into account the societal ramifications of their professions, maintain a competitive edge in the employment marketplace, and adhere to regulatory obligations. The omission of sustainability in educational curricula may result in future engineers and technical graduates being ill-equipped to contribute effectively to the attainment of SDGs. This could perpetuate unsustainable practices that have detrimental effects on both society and the environment. Hence, higher education institutions (HEIs) should prioritize student learning about sustainability such as the environmental ramifications of engineering projects, their social responsibilities, technological advances, and legal obligations ((To et al., 2023; Alenezi et al., 2023; Almawishir & Benlaria, 2024; Moharana et al., 2024). HEIs Could utilize digital technologies (DTs) such as AI, IoT and cloud computing to provide administrative, e-learning, and communication services (Hiran & Dadhich, 2024). Due to a lack of expertise in IT, limited access to hardware advancements, and challenges in updating software, higher education institutions in developing and underdeveloped countries may consider exploring cloud computing as a solution to provide high-quality education at a reasonable cost. According to UNESCO (2022), there are three main ways in which HEIs contribute to the SDGs: (a) by achieving their own targets related to SDG#4, which focuses on Quality Education; (b) by enhancing the standards and inclusiveness of the entire educational systems; and (c) by promoting the execution of the SDGs through innovative, multidisciplinary, and collaborative pedagogical, learning, and research methodologies (Embia et al., 2023a, 2023b).

The United Nations Organization's goals of poverty eradication, ensuring excellent education, and eliminating disparities have led to a growing attention on the association between education and sustainable development in recent years. Superior education is a basic right for everyone and is crucial for eliminating scarcity, promoting equal educational opportunities, and shaping inclusive learning environments, as stated in the SDGs. These facts provide the scholastic community with a new chance to cultivate efficient educational approaches that prioritize improving teaching and advancing education in a sustainable environment (Carrión-Martínez et al., 2020; Alshammari et al., 2023; Bonfield et al., 2020; Bauer et al., 2021). HEIs can advance and facilitate sustainable development (SD) and, subsequently, execution of the SDGs through several means. Firstly, by integrating sustainability ideologies into all the curricula and educational programs of the HEIs as well as in research endeavors (Leal Filho et al., 2024a, 2024b, 2024c). Secondly, by serving as crucial local centers of knowledge for sustainability. Lastly, by prioritizing sustainable development as a fundamental component in both planning and administrative processes (Leal Filho et al., 2024a, 2024b, 2024c). Therefore, HEIs have the potential to significantly contribute to the development of future leaders by actively participating in the transition towards a sustainable society. This can be achieved through the generation of knowledge and its dissemination to society, as well as by suitably educating students for their responsibilities in this society in the future (Mohamed Hashim et al., 2022; Busquets et al., 2021). Therefore, HEIs have the potential to significantly contribute in attaining the SDGs, while also reaping significant benefits from their involvement.

Though many initiatives have been taken in developed countries for enhancing the sustainability efforts in higher education sectors; these efforts are not that effective in developing and under developed nations (Akkaraboina et al., 2023). Hence, region specific studies are required to understand the challenges and prospects of implementing digital technology (DT) integrated sustainability initiatives of HEIs. The current investigation is performed with the following objectives:

- To understand the importance of incorporating sustainability into higher education system.
- To explore the opportunities in enhancing the sustainability initiatives of HEIs of Papua New Guinea (PNG) through the utilization of sophisticated digital technologies.

2 Research Methodology

A thorough literature study was undertaken to get a full understanding the role of digital technologies in sustainability practices in HEIs. The following scholarly databases were searched using specific search terms: ScienceDirect, IEEE Xplore, Scopus, and Web of Science. The search terms used included "digital technology," "Artificial Intelligence," "Internet of Things," "Information Communication Technology based higher education practices," "Disruptive technologies in higher education," and "sustainability practices in higher education sector" or "Sustainable development in higher education service providers". The inclusion criteria gave priority to peer-reviewed scholarly publications, conference proceedings, and industry reports that were published in the last five years (2019–2024). The studies that did not specifically focus on the higher education sector were not included in the analysis. The initial search produced 97 items. The exclusion criteria for this research include papers that are not journal articles, articles that are not available in full, and articles that are published in languages other than English. After a thorough evaluation based on the criteria of relevance, methodological rigor, and theoretical contribution, a final set of 25 publications were chosen for detailed examination (refer Fig. 1).

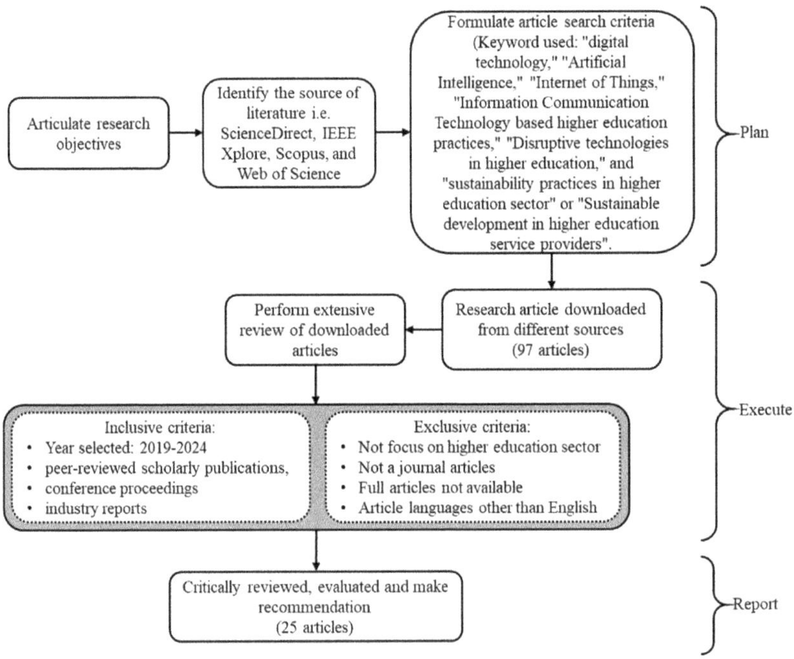

Fig. 1 Research methodology

3 Need of Digital Transformation in the System of Higher Education

In this section, the study explores potential digital technologies available to be utilized by the higher education sector and the associated benefits, refer Fig. 2. In recent years, there have been significant developments in educational technologies, which have resulted in creation of innovative tools and methodologies that are designed to improve the educational process. The use of analytics tools has provided teachers with the ability to monitor the progress of their students and tailor their teaching methods accordingly (Alshammari et al., 2023). By utilizing adaptive learning technologies, students are able to gain access to distinct educational opportunities that are suited to their specific requirements and preferred learning techniques. This has the potential to provide students with a more personalized educational experience (Galgal et al., 2023; Muduli & Barve, 2013a; Embia et al., 2023a; Idris, et al., 2024). Increasing the level of engagement and enjoyment in the learning process has been accomplished through the implementation of games and gamification, which has resulted in an increase in the level of student motivation and involvement. A more integrated and simplified learning environment has been created as a result of the internet of things, which has made it easier to integrate technology into everyday products. Students now have access to advanced technology and equipment thanks to the establishment

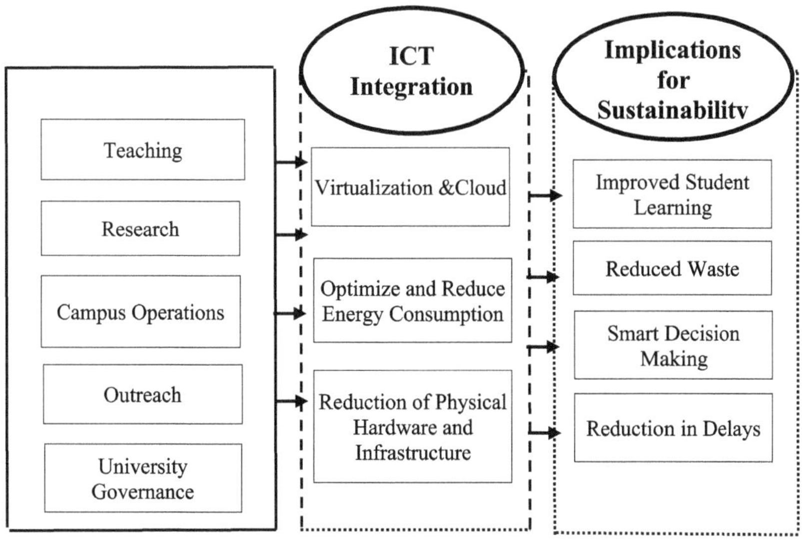

Fig. 2 Sustainable Development Framework for HEIs

of makerspaces, which were designed to give them opportunity to participate in activities that are creative and inventive. The traditional methods of instruction have been dramatically revolutionized by implementation of flipped classrooms, which prioritize students in the learning process. Wearable devices are being deployed to make it easier for teachers to track the level of involvement and progress of their students. According to Abad-Segura, et al. (2020), in order to cope with the changes that are brought through cutting-edge technologies, the digital transformation that has taken place in the education sector has necessitated the participation of sustainable leadership.

Implementation of digital technologies can enhance learning effectiveness in several ways(refer Fig. 3) and also contribute to sustainable development practices such as waste management. Reducing the use of papers, folders, biros or other items which can create solid waste that results in cleaner and greener environment. At the same time, it will be economical as work is done digitally which doesn't require additional costs for the purchasing of items that are likely to be replaced frequently. It also saves time and energy as most work will be done digitally. For example, big data makes it possible where course materials, textbooks, etc. can be customized and be accessible through digital technologies. Such digital technologies under artificial intelligence make it more convenient as you can have access to your work at finger tips. For example, laptops and smartphones are now portable that you can carry around and have access to your work. Therefore, digital technologies are sustainable in higher education despite challenges encountered which can be resolved strategically.

Many nations are working in accordance with the agenda for sustainable development in order to encourage advancements in the fields of education, information and

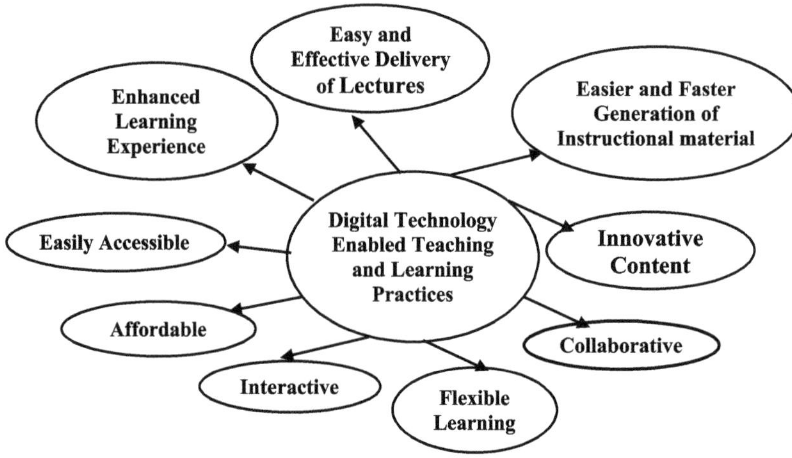

Fig. 3 Features of Digital Technology Enabled Teaching and Learning Practices (adopted from Haleem et al. [25])

communication technology, etc. and also included SDGs into their higher education programs. The higher education is essential for promoting sustainable development. HEIs are considered as facilitators of change that offer solutions to several critical issues facing countries, particularly in the area of sustainable perception and knowledge. Moreover, sustainability objectives can be realized by integrating information and communication technology in education and intelligent computing systems through the professional training by utilizing cutting-edge technologies in higher education institutions (Sady et al., 2019; Alahmari et al., 2019; Chernikova et al., 2020). Likewise, the advancement of intelligent computer technology and applications has resulted in substantial enhancements in education, pedagogy, and learning methodologies.

In a broader sense, higher education institutions are perceived to be key driver for achieving SDG targets. Therefore, all higher education institutions in all geography are placed in a responsible position to provide inclusive education for sustainable growth in response to SDG 4 achievement. Unfortunately, many higher education institutions in developing countries like Papua New Guinea, lack technological capability and also their approach towards addressing sustainability is a concern. Many discussions papers are available in this regard; however, the purpose of this study is to identify the various digital technologies that are available globally, which can be utilized in PNG higher education institutions. In general, this study identifies the following various categories as useful digital platforms that are available to transform higher education institutions.

3.1 Online Learning Platform

It's a leading platform available that are generally used for online for teaching and learning. It enables institutions to reduce their carbon footprint by minimizing travel and infrastructure requirements. Virtual classrooms, webinars, and e-learning modules are some of the sustainable practices.

3.2 Virtual Conference

Covid19 has forced virtual conference as alternative to physical gathering. Institutions can use this platform as it reduces travel - related virtual conferences reduce travel-related emissions, promote inclusivity, and allow for the dissemination of knowledge without geographical constraints.

3.3 Energy Management (EM) System

Higher education institutions are non-profit organizations and can implement intelligent energy management technologies to keep track and optimize consumption of energy across campus buildings. One example is smart lighting system is utilized sensors and control to adjust light levels based on utilization, daylight and time of day result in energy savings.

3.4 Renewable Energy (RE) Solution

Higher education institutions can integrate RE sources, such as solar panels or wind turbines, into their EM systems. These systems capture and utilize renewable energy, reducing reliance on fossil fuel-based electricity and lowering greenhouse gas emissions. This integration may fulfill the energy requirement for smart devices needed in teaching learning process.

3.5 Internet of Things (IoT) and Sensors

IoT devices and sensors can enhance learning and contribute sustainability. Instance smart classroom provides interactive whiteboards, smart projectors and other educational tools needed for education purposes.

3.6 Data Analytics

Institutions can utilize data analytics tools to drive sustainable development activities. It can be used to analyze and process many parameters viz. student behavior and waste management practices etc.

3.7 Collaboration and Communication Tools

Digital collaboration and communication tools enable students, faculty, and staff to work together on sustainability projects, irrespective of geographical boundaries. Online platforms, project management tools, and virtual collaboration spaces facilitate knowledge sharing, coordination, and teamwork.

3.8 Sustainable Apps

Higher education institutions can develop mobile applications that promote sustainable practices among students and staff. These apps can provide information on recycling, sustainable transportation options, local green initiatives, and sustainable lifestyle tips.

3.9 Virtual Laboratory and Simulation

The digital tools enable students to learn scientific concepts and conduct experiments in a virtual environment, reducing the environmental impact associated with traditional laboratory setups.

3.10 Sustainable Procurement System

Implementing digital procurement systems that consider sustainability criteria can help institutions make environmentally conscious purchasing decisions. These systems can assess suppliers based on their environmental performance, ethical sourcing, and carbon footprint, promoting sustainable supply chains.

4 Role of Digital Technology in Promoting Sustainable Development Practices in HEIs

Every human being has right to be educated and it is a fundament human right. Having to continue to a higher education is a privilege. Education plays a vital role in teaching necessary knowledge and skills to human beings. It also teaches attitudes and values which are required to contribute towards creating a sustainable future. Sustainability concerns have been integrated into curricula, policies and programs in the most colleges and universities in the United States. There is an organization called the Higher Education Sustainability Initiative (HESI) in which universities or higher education institutions can join. This organization is supporting sustainable development through higher education and has taken on board to organize High-Level Political Forum on sustainable development. HESI has come up with the initiative of "Education for Green Job" which was on improving education, career advising and job placement processes which was concerned on promoting green and sustainable career. This initiative started in 2021 and had its progressive report was out in 2022 with positive outcomes. The "StudentMedAID for International Need" is a charity from The University of Edinburgh in United Kingdom. The charity collects unwanted and outdated medical equipment and distributes abroad to under-resource medical institutions. By doing so, these items do not become waste but are put to good use.

4.1 Implementation of Digital Technology Integrated Sustainable Development Practices in Higher Education

Higher educational institutions are important as they can contribute in various perspectives in sustainable development, refer Fig. 4. In addition, to do so, there are five steps involve in carrying out a sustainability strategy. These steps include assessment, planning, implementation, evaluation and reassessment. Implementation of sustainable development practices requires critical planning in order to address it. The planning process should also take into consideration the three pillars of sustainable development which are social, economic and environment (Haleem et al., 2022; Chankseliani & McCowan, 2021; Mian et al., 2020). There has to be a vision that drives the plan towards achieving change for a better outcome for everyone. For instance, if improved waste management is the vision for a higher education institute, then plan may be to use digital technologies (Sahoo et al., 2021; Leal Filho et al., 2024a, 2024b, 2024c; Žalėnienė & Pereira, 2021). This has to be vision with clear plan. Then there will be a need to explore the options that may be necessary to achieve the SDG. You may have so many own ideas or ideas being shared by others but not every idea will be considered. The next thing is to decide on the best options to work on it. With your best options, there may also be need to do additional works such as having a detailed planning, costing as well as assessments for each of the options.

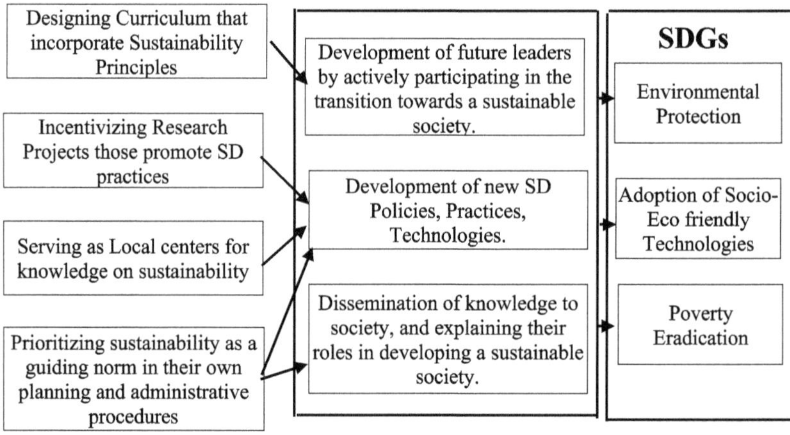

Fig. 4 HEI Strategy for Promoting SD Practices

When the best options have been selected, detailed planning comes into play. Each activity is planned according to its timeframe and budget. And the delegation of the task to the responsible people. For example, the current supplementary immunization activities (SIA) that has started last week for Morobe Province has a total budget of K500, 000.00. This has to be completed within a month and which requires nurses to reach child age 0 to 5 years old. A report will have to be submitted after these activities end to National Department of Health. After detailed planning, then it has to be reviewed and evaluated as to assess its impacts which can either be positive or negative depending on outcome it can be decided whether to continue on with what you have or to reconsider other better options to take. If the SIA program cannot reach its target indicated, then another option is to increase immunization vaccination in normal well baby clinics. Alternatively, to carry out more routine outreach immunization clinic throughout the year. Once you are satisfied with your proposed plan then it is time for implementation. When the plan has been accomplished, it is time to reassess the plan to identify areas of improvement and work on it in the future.

5 Implementation of Sustainable Development Practices in Higher Education Activities in PNG Employing Digital Technologies

5.1 Using AI in Higher Education

The world has gone through the digital technologies transition and is still continuing with new changes. Some countries are well off in terms of digital technologies whilst other countries are going at a snail's pace. The scenario in terms of digital technologies here in PNG may be different from other countries. Most of the populations live in the remote areas of PNG where there is no easy access to digital technology services. But those living in towns are exposed and are able to access these services. All higher educational institutions in PNG are located in towns meaning students and staffs are able to have access to digital technologies. There is a substantial level of incorporation of digital technology into the concept of teaching and learning in the academic world (Alenezi et al., 2023). Sustainable development practices can be implemented using AI in higher education (George & Wooden, 2023; Trevisan et al., 2024). The use of computers and smartphones can minimize the use of papers, folders and other items which contributes towards creating waste. By doing so, it improves waste management which will contribute towards a safe and friendlier environment. On the other hand, it saves time and makes work easier for the students (Tlili et al., 2023). However, there are or may be instances where not all students in higher educational institution are able to have a computer or even a smartphone. This can be a situation where the use of papers, folders and/or other items comes into play.

5.2 Using IoT in Higher Education

Developed countries have advanced more in digital technologies whilst developing countries are still trailing behind. In order to fill the gap, a quality sustainable education is required which can be made possible through the use of digital technologies such as IoT applications (Kuleto et al., 2021; Lozano et al., 2015). The use of IoT sensors can assist in identifying any danger that is about to happen and raise the alarm. By doing so, it prevents accidents and can also save lives. Smart boards in education have replaced texted books and blackboards. It has made teaching and learning more interactive and enjoyable. For example, it can clarify or explain complicated formula and even resolve students' queries in a shorter period of time. Students' attendance can also be monitored through IoT base attendance system. Keeping accurate attendance of students can also save costs as well as encouraging students to attend all classes. The IoT powered screen readers can now help students with disabilities that are having difficulty in learning to be able to learn just like a normal student. For instance, the IoT powered screen can turn spoken words into written words for them to understand.

5.3 Using Big Data in Higher Education

Big data is very important in higher education as well as for sustainable development practices. Through big data, institutions may be able to plan for the future. For example, data collected and analyzed can be used to make projections on the enrollment of new students or the use of resources. The institution can also plan for their specific programs and degrees. Academic staff can be empowered through the use of big data. Customized learning resources such as textbooks, e-books, online resources and school software programs can be create through big data. By lessening the work load for academic staff, it gives them more time in classroom attendance. For example, tests and assignments will be done automatically using the learning management system. Big data creates new opportunities. Digital innovations and technologies are able to improve students' engagement, outcomes and experiences. The 5G cellular network and block chain are the future of big data technology. This is to ease the internet traffic, keeping students' data secure and enhanced mobile data abilities.

6 Opportunities in Using Digital Technology

Digital technologies have many opportunities that higher education can tap into, refer Table 1. ***Digital learning technology*** is one application where lecturers or instructors can use to facilitate learning. The course resources can be made available on this application for students to have access to it. ***Instructional modalities*** are the ways or avenues in which learning is taking place. An example would be on campus technology enhanced. This is when students attend in person but their instructions are given through the use of technology. ***Personal support services*** are also required for a successful use of digital technologies. ***Organizational policies and planning*** are required in valuing digital teaching and innovations. ***Instructor development*** is vital as the instructors have to be equipped with the knowledge and skills in delivering online teaching. ***Learners' development*** is an opportunity they get as it prepares them to enter into the workforce. What they have learnt may be applicable and the knowledge and skills gained will become useful in the actually workforce. ***Partnership*** also plays an important role in online teaching and learning. For example, professional organizations that provide digital teaching may be able to provide trainings or resources for the online teaching and learning for the institution.

Table 1 Digital Technology Based Teaching Learning Opportunities in PNG

Opportunities	Description
Digital learning technology	One application where lecturers or instructors can use to facilitate learning
Instructional modality	Ways or avenues in which learning is taking place
Personal support services	Support services required for a successful use of digital technologies
Organizational policies and planning	Required in valuing digital teaching and innovations
Instructor development	Instructors have to be equipped with the knowledge and skills in delivering online teaching
Learners' development	An opportunity they get as it prepares them to enter into the workforce
Partnership	Professional organizations providing digital teaching may be able to provide trainings or resources for online teaching and learning for the institution.

7 Challenges Encountered Using Digital Technology

Newer processes and technologies frequently encounter resistance to change from long-term employees who believe there is nothing wrong with the way things have traditionally executed. It can be challenging to establish a distinct digital transformation plan that fits with an organization's overall strategic goals, and progress can be slowed down by a lack of vision or leadership. Although it may appear unambiguous, it is important to reiterate: digital transformation does not entail substituting humans with technology, but rather enhancing human capabilities to collaborate more effectively with technology. Digital transformation must prioritize human elements, achievable just with the appropriate personnel.

The challenges encountered by higher educational institution in digital technologies are data security. It is costly for an institution improve security network which means that students' data may be at stake. There should be IT support for both students and staff that should be captured by the IT strategy. Responsible administrator should seriously look into it. There is a demand on IT staff in most organizations. In order to meet the demand, there must be funds available for the staff who will be meeting the demand. For example, if an IT person is taken on board to work, he has to be paid or compensated for the work s/he is doing. Without digital integration there won't be efficiency in IT. This will lead to inefficiencies as well as redundancies. Data governance is a big challenge for higher educational institution. There should be an officer to manage data for specific institution. Adapting to new changes is often a challenge for people within an organization. The changes taking place should be supported by all partners and stakeholders within the organization.

8 Recommendations

The scope of this research is limited; however, this study still has the potential to be expanded or direct future research. Therefore, the following recommendations appropriate for academics, researchers, Governments, institutions and individual.

1. There is a great need for conception of SDGS in HEI for driving force to achieve the agenda 2030. Globally, many institutions in developed countries are taking lead but limited drive in the developing countries like PNG. Therefore, there is a need for deep qualitative research in developing a mechanism for higher education institutions to integrate SDG goals into their practices. For example, in PNG a questionnaire can be send to universities and colleges with clear research objective to determine the lack of drive-in implementing SDG goals.
2. As mentioned, the need model developed by this research is biased; however, the conceptual mode is unique and thus, requires a review and deeper qualitative and quantitative research. A broader scope is recommended and it will require advance research method for empirical testing to determine the validity of the model.
3. A general perspective to this research indicated the concept of sustainable practice missing in the higher education institutions in developing countries particularly in PNG. Therefore, there is a need to review the policies of the Department of Higher Education, Science, Research and Technology Act to include SDG targets in relation Education for Sustainable Development.
4. Data security can be addressed using guidelines from cyber security framework. Such framework may ensure that there is protection within the IT environment of the institution. For an institution to support its students and staff it has to learn which tools are practical and realistic and that their budget can cater for is something that the university or institution have to know. In addition, how those tools can be combined
5. The need for IT staff is a concern in most organization because it is an area of expertise that is on demand. Worthwhile strategies have to be considered to address this concern such as conduct cross-training for staff within the organization. Improve recruitment processes for IT staff and as much as possible sustainable IT workforce.
6. Digital integration will need IT experts to create a continuous architecture of IT network which will improve digital integrations.
7. Data governance can be seen as a long-term sustainable process. It starts off with smaller work before leading up to bigger data sets. This means information is categorized before sorted out for the use of business intelligence.
8. There are couple of ways in which adaptation to change can be facilitated. This includes developing policies which supports learning analytics and keeping uniform standards for data service providers to abide by. New technology practices to consider and to offer pedagogical trainings.

9 Implications of the Study

This research has not only significant theoretical implications but also has certain practical implications which are discussed in following sub-sections.

9.1 Implications for Theory

This research contributes to the technology integration theory (Consoli et al. 2023), that explores how integration of technology to traditional operational practices could enhance performance. This research reveals that technology integration to operational practices of higher educational institutions could be helpful in improving their sustainable development practices. This research proposed a framework 'Sustainable Development Framework' for HEIs which helps in visualizing the areas where technology integration could be helpful in enhancing SD performance of HEIs. Our research contributes to 'SDG 4' which emphasizes on provision of 'quality education', 'SDG 11' that stresses on 'sustainable cities and communities'.

9.2 Practical Implications

Sustainability in Higher Education through implementation of advanced digital technologies is the way forward in today's higher education. It is focused around sustainability as well as being inclusive with its products. For example, the IoT powered screen reader makes it easier by turning spoken words into written words. This research has explored how HEIs can contribute to SDGs of any nation and help building a sustainable society. It also proposed a conceptual framework to portray how the digital technologies can help HEIs in their effort of building sustainable societies and nations. This can help decision makers to develop suitable strategies to encourage higher involvement of HEIs in sustainable development programs of the countries especially the emerging economies. This research further, revealed several opportunities and challenges for HEIs while developing sustainable digital transformation framework (SDTF) and implementing the same. It also recommended HEIs of PNGs the action plans for effective implementation of SDTF.

10 Conclusion

To conclude, the world has been and is now centered on digital technologies to get work done. Sustainable development practices are also dependent on digital technologies in order to achieve the sustainable development goals. For instance, sustainability in higher education through implementation of digital technologies. There are always opportunities and challenges in developments. This is the same with digital technologies when implemented in the higher education. However, the opportunities are overwhelming as digital technologies have advanced in the current times. Most work is done through digital technologies. Using artificial intelligence, internet of things and big data applications, making work easier and faster as well as saving time, costs and energy. At the same time the digital technologies are now becoming inclusive for people with disabilities. Widespread application of digital technologies and their benefits attracted universities to integrate these technologies to stream line their operations. This factor motivated us to explore how this integration contributes to the sustainable development practices. Realizing the fact that the advancement of any field is dependent upon the establishment of a rational fusion of prior inquiries and their findings (Muduli et al., 2016; Bihari et al., 2022; Peter et al., 2022) this study adopted literature review approach to explore the opportunities for sustainability in higher education sector through employment of digital technologies. The literature review approach has been advocated by many of researchers owing to the fact that it has the potential to make a significant contribution to the creation of methodological, thematic, and conceptual frameworks in a variety of fields (Dhingra et al., 2024; Swain et al., 2021). Hence, this study is conducted through an extensive review of pertinent literature. However, this could be also considered as one of the limitations as it does not provide an idea about the nature and direction of interrelationships existing among various elements influencing the system. In future studies the factors influencing sustainable digital transformation (SDT) of HEI will be explored and their influence would be quantified through analytical hierarchy process (Muduli et al., 2013a, 2013b; Embia et al., 2023a, 2023b) or graph theoretic and matrix approach. Evaluation of degree of influence of various factors on SDT will help decision makers in formulating strategies for reducing their negative impacts.

References

Abad-Segura, E., González-Zamar, M. D., Infante-Moro, J. C., & Ruipérez García, G. (2020). Sustainable management of digital transformation in higher education: Global research trends. *Sustainability, 12*(5), 2107.

Akkaraboina, M., Kodari, S. R., Barve, A., & Muduli, K. (2023). Securing data of online learners using Blockchain technology. In *Advancements in artificial intelligence, Blockchain technology, and IoT in higher education* (pp. 181–202). Apple Academic Press.

Alahmari, M., Issa, T., Issa, T., & Nau, S. Z. (2019). Faculty awareness of the economic and environmental benefits of augmented reality for sustainability in Saudi Arabian universities. *Journal of Cleaner Production, 226*, 259–269.

Alenezi, M., Wardat, S., & Akour, M. (2023). The need of integrating digital education in higher education: Challenges and opportunities. *Sustainability, 15*(6), 4782.

Almawishir, N. F. S., & Benlaria, H. (2024). Innovation and dimensions of sustainable development: Mediating role of ICTs. *Cogent Business & Management, 11*(1), 2336306.

Alshammari, A. M., Alshammari, F. F., Thomran, M., & Altwaiji, M. (2023). Integrating technological knowledge into higher education curricula: An essential measure for attaining sustainable development in Saudi Arabia. *Sustainability, 15*(22), 15956.

Bauer, M., Rieckmann, M., Niedlich, S., & Bormann, I. (2021). Sustainability governance at higher education institutions: Equipped to transform? *Frontiers in Sustainability, 2*, 640458.

Bihari, A., Dash, M., Kar, S. K., Muduli, K., Kumar, A., & Luthra, S. (2022). Exploring behavioural bias affecting investment decision-making: A network cluster based conceptual analysis for future research. *International Journal of Industrial Engineering and Operations Management, 4*(1/2), 19–43.

Bonfield, C. A., Salter, M., Longmuir, A., Benson, M., & Adachi, C. (2020). Transformation or evolution? Education 4.0, teaching and learning in the digital age. *Higher education pedagogies, 5*(1), 223–246.

Busquets, P., Segalas, J., Gomera, A., Antúnez, M., Ruiz-Morales, J., Albareda-Tiana, S., & Miñano, R. (2021). Sustainability education in the Spanish higher education system: Faculty practice, concerns and needs. *Sustainability, 13*(15), 8389.

Carrión-Martínez, J. J., Luque-de la Rosa, A., Fernandez-Cerero, J., & Montenegro-Rueda, M. (2020). Information and communications technologies (ICTs) in education for sustainable development: A bibliographic review. *Sustainability, 12*(8), 3288.

Chankseliani, M., & McCowan, T. (2021). Higher education and the sustainable development goals. *Higher Education, 81*(1), 1–8.

Chernikova, O., Heitzmann, N., Stadler, M., Holzberger, D., Seidel, T., & Fischer, F. (2020). Simulation-based learning in higher education: A meta-analysis. *Review of Educational Research, 90*, 499–541.

Consoli, T., Désiron, J., & Cattaneo, A. (2023). What is technology integration and how is it measured in K-12 education? A systematic review of survey instruments from 2010 to 2021. *Computers & Education, 197*, 104742.

Dhingra, S., Raut, R., Naik, K., & Muduli, K. (2024). *Blockchain technology applications in healthcare supply chains–a review*. IEEE Access.

Embia, G. J., Moharana, B. R., Behera, B. C., Mohmaed, N. H., Biswal, D. K., & Muduli, K. (2023a). Reliability prediction using machine learning approach. In *Smart Technologies for Improved Performance of manufacturing systems and services* (pp. 21–37). CRC Press.

Embia, G., Mohamed, A., Moharana, B. R., & Muduli, K. (2023b). Edge Computing-Based Conditional Monitoring. Intelligent Manufacturing Management Systems: *Operational Applications of Evolutionary Digital Technologies in Mechanical and Industrial Engineering*, 249–270.

Galgal, K. N., Ray, M., Moharana, B. R., Behera, B. C., & Muduli, K. (2023). Quality control in the era of IoT and automation in the context of developing nations. In *Smart Technologies for Improved Performance of manufacturing systems and services* (pp. 39–50). CRC Press.

George, B., & Wooden, O. (2023). Managing the strategic transformation of higher education through artificial intelligence. *Administrative Sciences, 13*(9), 196.

Haleem, A., Javaid, M., Qadri, M. A., & Suman, R. (2022). Understanding the role of digital technologies in education: A review. *Sustainable Operations and Computers, 3*, 275–285.

Hiran, K. K., & Dadhich, M. (2024). Predicting the core determinants of cloud-edge computing adoption (CECA) for sustainable development in the higher education institutions of Africa: A high order SEM-ANN analytical approach. *Technological Forecasting and Social Change, 199*, 122979.

Idris, M. D., Feng, X., & Dyo, V. (2024). *Revolutionising higher education: Unleashing the potential of large language models for strategic transformation*. IEEE Access.

Kuleto, V., Ilić, M., Dumangiu, M., Ranković, M., Martins, O. M., Păun, D., & Mihoreanu, L. (2021). Exploring opportunities and challenges of artificial intelligence and machine learning in higher education institutions. *Sustainability, 13*(18), 10424.

Leal Filho, W., Eustachio, J. H. P. P., Nita, A. C., Dinis, M. A. P., Salvia, A. L., Cotton, D. R., & Dibbern, T. (2024a). Using data science for sustainable development in higher education. *Sustainable Development, 32*(1), 15–28.

Leal Filho, W., Lange Salvia, A., Beynaghi, A., Fritzen, B., Ulisses, A., Avila, L. V., & Nikolaou, I. (2024b). Digital transformation and sustainable development in higher education in a post-pandemic world. *International Journal of Sustainable Development & World Ecology, 31*(1), 108–123.

Leal Filho, W., Ribeiro, P. C. C., Mazutti, J., Lange Salvia, A., Bonato Marcolin, C., Lima Silva Borsatto, J. M., & Viera Trevisan, L. (2024c). Using artificial intelligence to implement the UN sustainable development goals at higher education institutions. *International Journal of Sustainable Development & World Ecology*, 1–20.

Liu, J., Kitamura, Y., & Savelyeva, T. (2022). Building an 'ecosystem'for transforming higher education teaching and learning for sustainability. *Asia Pacific Education Review, 23*(4), 539–542.

Lozano, R., Ceulemans, K., Alonso-Almeida, M., Huisingh, D., Lozano, F. J., Waas, T., & Hugé, J. (2015). A review of commitment and implementation of sustainable development in higher education: Results from a worldwide survey. *Journal of Cleaner Production, 108*, 1–18.

Mian, S. H., Salah, B., Ameen, W., Moiduddin, K., & Alkhalefah, H. (2020). Adapting universities for sustainability education in industry 4.0: Channel of challenges and opportunities. *Sustainability, 12*(15), 6100.

Mohamed Hashim, M. A., Tlemsani, I., & Duncan Matthews, R. (2022). A sustainable university: Digital transformation and beyond. *Education and Information Technologies, 27*(7), 8961–8996.

Moharana, B. R., Behera, B. C., & Muduli, K. (Eds.). (2024). *Digital technology enabled circular economy: Models for environmental and resource sustainability*. CRC Press.

Muduli, K., & Barve, A. (2013a). Developing a framework for study of GSCM criteria in Indian mining industries. *APCBEE Procedia, 5*, 22–26.

Muduli, K., & Barve, A. (2013b). Modelling the behavioural factors of green supply chain management implementation in mining industries in Indian scenario. *Asian Journal of Management Science and Applications, 1*(1), 26–49.

Muduli, K., Barve, A., Tripathy, S., & Biswal, J. N. (2016). Green practices adopted by the mining supply chains in India: A case study. *International Journal of Environment And Sustainable Development, 15*(2), 159–182.

Peter, O., Swain, S., Muduli, K., & Ramasamy, A. (2022). IoT in combating COVID-19 pandemics: Lessons for developing countries. In *Assessing COVID-19 and other pandemics and epidemics using computational modelling and data analysis* (pp. 113–131). Springer.

Sady, M., Żak, A., & Rzepka, K. (2019). The role of universities in sustainability-oriented competencies development: Insights from an empirical study on polish universities. *Administrative Sciences, 9*, 62.

Sahoo, K. K., Muduli, K. K., Luhach, A. K., & Poonia, R. C. (2021). Pandemic COVID-19: An empirical analysis of impact on Indian higher education system. *Journal of Statistics and Management Systems, 24*(2), 341–355.

Swain, S., Peter, O., Adimuthu, R., & Muduli, K. (2021). Blockchain technology for limiting the impact of pandemic: Challenges and prospects. Computational modeling and data analysis in COVID-19 research, 165–186.

Tlili, A., Nascimbeni, F., Burgos, D., Zhang, X., Huang, R., & Chang, T. W. (2023). The evolution of sustainability models for open educational resources: Insights from the literature and experts. *Interactive Learning Environments, 31*(3), 1421–1436.

To, T. T., Al Mahmud, A., & Ranscombe, C. (2023). Teaching sustainability using 3D printing in engineering education: An observational study. *Sustainability, 15*(9), 7470.

Trevisan, L. V., Eustachio, J. H. P. P., Dias, B. G., Filho, W. L., & Pedrozo, E. Á. (2024). Digital transformation towards sustainability in higher education: State-of-the-art and future research insights. *Environment, Development and Sustainability, 26*(2), 2789–2810.

United Nations Organizations. 2015. The 17 Sustainable Development Goals. (Retrieved on 5 February 2022 from). https//www.un.org/sustainabledevelopment/es/objetivos-de-desarrollo-sostenible/

Žalėnienė, I., & Pereira, P. (2021). Higher education for sustainability: A global perspective. *Geography and Sustainability, 2*(2), 99–106.

Challenges and Ethical Considerations in Integrating AI into Higher Education

Padmesh Tirunelveli Narayanapillai, Thanammal Indu Vijayalakshmi, and Velu Vengadeshwaran

Abstract Efforts have been made in defining artificial intelligence in higher education increasingly as advocacy relationships in shifting paradigms in the methods of teaching, learning, and administration to develop a personalized learning experience, improve decision-making, and operation efficiencies. These transformations are in the aspect of alignment with the United Nations Sustainable Development Goals, especially about SDG4 on "inclusive and equitable quality education". Unfortunately, such developments may also bring possible challenges and ethical issues relevant to their adoption. Some of these possible challenges include guaranteeing data privacy and security in relation to a possible dependence on data sets, tackling algorithmic biases that would threaten equity, and improving the overall technology gap to avoid more resource-poor institutions from being further marginalized. New pedagogical challenges arise as the possible revolution comes with the integration of AI-driven tools in the classroom, balancing all the benefits technology may bring with principles about what teaching at its best entails. Scalability and resource limitations add to disparity in distribution of the implementation. Issues relating to ethics also have considerable resonance, directly relevant to accountability and transparency of AI systems, the handling of intellectual properties, and cultural sensitivity to include a lot more. These issues further involve the ethical dimension of automations, their addressing with job insecurities concerning retaining student autonomy and critical thinking with the environment-involved AI production. This requires strong policy frameworks, initiatives to build capacities, interdisciplinary collaboration, and a good foundation in ethical principles for AI design. Therefore, it will inspire stakeholder

P. Tirunelveli Narayanapillai (✉)
School of Engineering, Asia Pacific University of Technology and Innovations, Kuala Lumpur, Malaysia
e-mail: drtvnpadmesh@gmail.com

T. I. Vijayalakshmi
Department of Computer Science and Engineering, Amrita School of Computing, Amrita Vishwa Vidyapeetham, Nagercoil, Tamil Nadu, India
e-mail: thanammal.indu@gmail.com

V. Vengadeshwaran
School of Engineering and Computing, MILA University, Nilai, Malaysia
e-mail: vengadeshwaran.velu@mila.edu.my

A. Ara (ed.), *Role of Artificial Intelligence in Higher Education*,
https://doi.org/10.1007/978-981-95-5350-1_5

involvement in an AI lifecycle as a fusion of trust and technology to educational goals and values. Addressing the above challenges within the AI system will immediately convey the importance of high participation among industrial stakeholders.

Keyword Artificial Intelligence · Ethical Challenges · Data Privacy · Algorithmic Bias · Equity Pedagogical Innovation · Policy Frameworks · Interdisciplinary Collaboration

1 Introduction

The fast-increasing Artificial Intelligence infusion into all its varied applications has transformed the higher education scenario through teaching, learning, and administrative processes, offering huge potential in enhancing educational experiences, from creating pathways for personalized learning to informed decision-making and operational efficiency. Adaptive learning systems, intelligent tutoring platforms, and AI-enabled analytics can help teachers address diverse learning needs and possibly optimize resource allocation, along with better evidence of student progress-a practice that closely relates to the United Nations SDG 4, "Ensure inclusive and equitable quality education". Through the use of virtual reality and augmented reality applications, artificial intelligence (AI) has the potential to enhance the educational process by enabling the development of immersive learning environments (Eden 2024).

On the other hand, the integration of Artificial Intelligence into higher education should not be blind to several challenges and related ethical issues it comes with. Most of the challenges are with data privacy and security concerns of the AI system as these must deal with huge data sets that depend on students' sensitive information. Other threats include algorithmic biasedness, which is often related to higher education institutions, especially in the realm of admission, grading, and assessment. When developing algorithms, studying data sets, and making AI decisions, gender bias must be avoided. To guarantee traceability and transparency, the best possible data collection, labeling, and algorithm documentation are required (Slimi 2023). Inequitable data involved may become a cause of playing against disadvantages among groups that stay underrepresented. There is a foreseen digital divide between high and low resource institutions which is very likely to develop huge educational inequalities and limit the access of the AI-powered tool to marginalized societies.

Pedagogically, in overdependence on AI, it would undermine the humanness of pedagogy, critical thinking, creativity, and empathy amongst many others. Furthermore, ethical issues further complicate the use of AI in availing learners with the definite required accountability, transparency, and finally knowledge rights. Besides that, job security and sensitivity to diversity in designing automated tools are yet other causes for concern.

In view of such challenges, it therefore becomes necessary to have in place solid policy frameworks that will enhance ethical principles: fairness, accountability, and

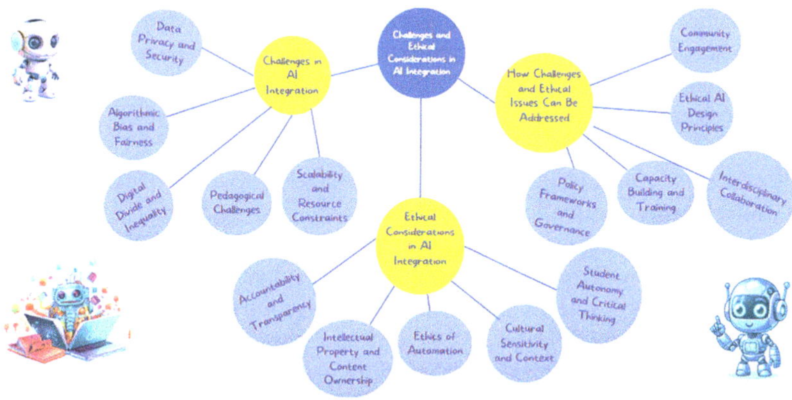

Fig. 1. Challenges and ethics of AI integration in higher education

transparency. The availability of this capacity, intervention mechanisms, multidisciplinary collaboration, and the active participation of all stakeholders will make the systems of AI relevant in respect of the attainment and value of education. Only when the educational institutions can instill confidence and inclusiveness among them can this technology be tapped into for responsible utilization and help make education an enduring, equitable, and effective outcome. Fig. 1. shows the challenges and ethics of AI integration in higher education framework.

2 Challenges of AI Integration

Integrating Artificial Intelligence (AI) into higher education is poised to cause transformative possibilities within higher education, radically altering how teaching and learning processes are administrated and even improving the effectiveness of institutional actions. One of the biggest challenges to integrating AI in education is making sure that all students, irrespective of their socioeconomic background or geographic location, have equitable access to AI-powered tools and resources (Eden 2024). Indeed, AI-powered systems offer personalized learning experiences, optimize decision making, and improve efficiency-exploring United Nations Sustainable Development Goal (SDG) 4, which encourages inclusive and equitable quality education. However, the integrated AI systems are not free from challenges, for these advances show all sorts of dismal scenarios that might need to be cured for fair and responsible implementation. The studies indicate that training both the faculty and the staff for an AI adoption has affected a significant barrier. Professional growth opportunities along with strong support networks for the successful absorption of AI technologies is paramount (Saaida, 2023).

Data privacy and security are important issues because those often require submission of sensitive data about students in large amounts, thereby posing a significant data breach and misuse risk. Algorithmic bias makes matters worse. Any AI system built using data biased along certain dimensions can put equity at threat in terms of admissions and grading. The digital divide between high-resource and low-resource institutions also increases inequality about the access to AI technologies in underserved areas.

There are also pedagogical issues whereby teachers will have to take on different roles with the ensuing technology, which tends to diminish the human part of teaching that evokes critical thinking and empathy. Scalability and resource constraints such as those faced by low-income regions also have a negative impact on the adoption of AI. The application of AI technologies in higher education is very much multifaceted and extremely expensive (Saaida, 2023). It is necessary to have strong frameworks coupled with equal access and inter-disciplinary collaboration to fully realize the potentials of AI and at the same time minimize the disasters with which it is associated while transforming education.

2.1 Data privacy and Security

The adoption of Artificial Intelligence (AI) applications in higher education is promising when it comes to personalized learning, efficient administration, and improved decision-making; however, it also raises significant challenges, especially in the domain of data privacy and security. Educational institutions and technology providers grapple with a range of broad legal and ethical issues around the collection, use, and sharing of sensitive student data and compliance by such institutions with the laws such as the Family Educational Rights and Privacy Act (FERPA) in the USA (Eden 2024). This is so because the AI systems require considerable datasets that typically contain sensitive information like metrics on student performance and how to access demographic information, as well as behavioral patterns, so that very serious issues on the part of data handling come into play with respect to breaches, misuse, and conformity with stringent data protection laws such as the General Data Protection Regulation (GDPR).

Data Privacy Issues. Thus, the implication of anything that goes wrong—given the inclusion of personally identifiable information (PII) in higher education datasets—is severely identity theft, monetary losses, or reputational damage to the institutions. To ensure privacy, this would involve strong encryption of access to the database and conducting regular audits on security loopholes. Institutions should also be sensitized stakeholders on the importance of data privacy and their roles in this enterprise

Data Security Challenges. Among other things, protection against system breaches will keep the AI system secure and, thus, safeguard the integrity of an educational process. Attacks, for instance, cyber-hacking or phishing, or malware attacks, also form a strong threat to the vast AI processed data. The kind of measures required to achieve this include comprehensive cybersecurity measures, such as firewalls, intrusion detection systems, and repetitive security assessments. Moreover, incident response plans would be a step toward reducing the negative impact of security events by responding timely to breaches.

GDPR Compliant. The conditions in GDPR related to data protection are very stringent when it comes to user consent, data minimization, the right to be forgotten, etc. Higher education institutions are required to ensure such alignments with their AI systems so as to avoid staggering penalties and litigations by ensuring Data Protection Impact Assessments (DPIAs), Data Protection Officers (DPOs), and transparent data processing activities. Complying with GDPR not only makes an institution legally compliant but also develops confidence among the students and other stakeholders.

Building Trust Among Stakeholders. One of the most significant ways in which trust will be a major pillar for the adoption of AI in higher education is openness: how institutions will be open in their AI systems' collection, processing, and usage of students' information. This should also include good communication channels and involve stakeholders, faculty, and students in decision-making to bring them into the trust culture. Another factor is bringing visibility to feedback mechanisms and presenting rapid interventions for concerns.

Best Practices for Data Protection and Privacy Security. Here are some best practices that the universities can opt for the gray areas in the data to work through—data privacy and security.

- Data Minimization: It collects only such data that is necessary for AI functions; the scope within which data breach exposure might occur should be minimized.
- Anonymity: The anonymity and secrecy provisions shall apply wherever robust encryption is used.
- Regular Audits and Assessments: This includes regular security audits and risk assessments to identify and mitigate vulnerabilities.
- Stakeholder education: Educating students, faculty, and staff about the principles of data privacy and security.
- Policy document: Document clear policies on data collection, storage, processing, and sharing.
- Incident Response Planning: Derive and regularly test updated incident response plans, which should cover effective needs in incidents concerning data breaches and security.

AI can hugely transform the space of higher education, but it will not be done without answering the many challenges in data privacy and security. Only with proper safeguards in place will the full benefits of AI be adopted by institutions without breaching the rights and privacy of people, coupled with the compliance of institutions with data protection regulation policies like GDPR through an open line of communication with their stakeholders.

2.2 Algorithmic Bias and Fairness

The algorithmic bias in Artificial Intelligence (AI) is really a challenge in its fairness and equity as far as higher education is concerned. In fact, most of these biases are the result of training data that have included the existing social biases within society or coming from designs that usually unintentionality favors a group over other groups. For example, AI-enabled admission systems may be skewed to students from wealthy backgrounds because historical data is most often biased towards that demographic, hence depriving diversity and inclusivity. AI algorithms, which base their claims entirely on specific data that draw different lines in the sky, end up missing out on real social intersections. Such paradigms do not even allow healthy rethinking of the production of evidence where they have been brought directly to service delivery and social justice sensitive settings as a social construct (Agarwal 2023).

Understanding Algorithmic Bias. Algorithmic prejudice indicates that an AI system would produce systematically prejudiced results for a given situation or subject because of some incorrect assumptions of machine learning. This is reflected in educational contexts where students cannot experience user-friendly environmental conditions owing to race, gender, socioeconomic background, etc. Academics have raised eyebrows for many years about the potential disparity of effectiveness and generalizability across populations of education algorithms. Most publicized this debate with the 2020 UK General Certificate of Secondary Education (GCSE) and A-Level grading controversy, where a set of formulas was developed by the national qualifications regulator (by hand rather than by an automated algorithm) to assign predicted examination grades based on teacher predictions, whereas students from state-funded schools had poorer grades awarded by the algorithm when compared to better grades (even better than teacher prediction) for students from smaller independent schools (Baker 2022).

Effects on Higher Education. Thus, it means that admission systems with a bias in AI would instead compound the already existing inequality and worsen the exclusion of these students. It would also cause damage to the credibility of such systems while underlining that trust in education institutions in terms of diversity and inclusion becomes eroded.

Mitigating Algorithmic Bias. Addressing algorithmic biases must be approached multifariously:

- Diverse and Representative Data: using diverse demographic representation in the training dataset would enable the AI systems to learn equitable patterns. The lack of diversity in training data is one source of biased AI admissions models.
- Detection and Correction of Bias: Adoption of tools that will expose and make bias in AI algorithms detectable must be crucial. Fairness-aware Machine Learning techniques will ensure fairness in decision making.
- Ethical AI Frameworks: Development and compliance with the ethical guidelines must ensure that the design and deployment of the systems have an inherent fairness factor. These would include the process by which the decisions are made by the AI, their transparency and the accountability for the results produced.

- Humans: Makes possible the situation where humans are still part of the decision making in AI so that all biased decisions can be evaluated and creatively intervention. Human judgment would be enhanced instead by AI, and this would fall into sensitive issues such as admissions related issues.

Case study: Using Artificial Intelligence in the College Admission Process. AI can either lessen or amplify biases in college admissions. Properly strategically using AI will systematically analyze patterns to help find and rectify inconsistencies and thus foster justice. In contrast, however, biased data trained on an AI system will reinforce the present differentials. Experts concede that artificial intelligence has the potential to solve equity gaps and, at one and the same time, increases them, depending on the use of these tools.

Artificial intelligence can indeed improve decision-making in higher education; however, constant vigilance must be exercised to prevent algorithmic biases from detracting from fairness and equity. Schools will have to develop mechanisms to avoid discrimination in the power that AI would give them while safeguards are put in place on the inputs through diverse data, bias detection mechanisms, ethical frameworks, and human oversight.

2.3 Algorithmic Bias and Fairness

In educational institutions, the newly introduced Artificial Intelligence (AI) will benefit productively personalized learning and better the learning experience. Still, such new introductions into the education system further strengthen the present disparities owing to the digital divide effects through which differences arise in access regarding technology and the internet. For instance, students and learning institutions from the less resourceful settings would be able to afford appropriate Wi-Fi access and meaningfully engage in AI, making it difficult to attain equity in achieving world distribution as far as access to education is concerned.

Understanding the Digital Divide. This term differentiates those who possess modern information and communication technology from those who are not included because they do not have access to the above-mentioned technology. In education concerning AI, this digital divide manifests itself in the following ways:

- Deficits in infrastructure: This would include high-speed internet; adequate hardware, as well as reliable sources of electricity, which are essential for deploying AI technologies in most low-resource institutions. For example: students residing in areas remote and isolated from the city's pulse will find it difficult to access much-needed AI-imparted learning because there is an unreliable connection to the internet.
- Gaps in Educational Resources: Even if most of the infrastructure is available, most underserved students have inadequate educational resources, such as qualified teachers or high-quality curriculum, needed to support their AI literacy. Without proper AI literacy, the AI tools utilized by students in under-resourced

schools will likely be less-effective as compared to the usage of the same tools by students in resource-rich schools.

Impacts on Education Equality. This digital divide, when accompanied by a modernization of artificial intelligence, can lead to increased inequalities across the following levels:

- Worsening of the Socioeconomic Gap: The richer students would be able to afford awesome AI tools to help them learn better even as they continue widening the gap with their less privileged peers.
- Limited AI Proficiency: Children from poorly endowed resource devices would lack the proper skills in the future because of the absence of exposure to AI technologies, ultimately, these children will be left out of jobs in an industry that will be increasingly driven by AI in the future.
- Differences in Educational Outcomes: The benefits of AI, such as personalized learning and efficient resource management, are likely to be enjoyed by a few resource-rich institutions.

Bridging the Digital Divide. Possible measures should include the following:

- Infrastructure developments. Installation of computers to underserved areas along with reliable internet access and electricity development would indeed do much towards the purpose. A public-private partnership could go a long way in making this.
- Replacement of the Digital Divide: AI Literacy Programs- Formal education efforts would build AI literacy skills in lower-resource settings to prepare students and their teachers to harness the power of AI tools. For example, UNESCO has advocated actions for the promotion of AI literacy to bridge the AI divide.
- Policy Interventions: Policy formulations in governments and educational agencies could enhance accessibility to AI technologies such that marginalized communities will not be cut out.
- Open Educational Resources (OER): The creation and dissemination of high-quality free educational materials in AI possess the kind of potential that could open AI to all learners outside of their socioeconomic status.

AI has much potential to rework education; however, there is a need for tackling the digital divide to prevent widening existing inequalities. Stakeholders should hence work towards building an environment of all-inclusive education, with access to all benefits of AI, through infrastructure development, provision of AI literacy, policies interventions, and open educational resources.

2.4 Pedagogical Challenges

AI in education can afford incredible possibilities as well as great barriers to pedagogy. Intelligent tutoring systems and adaptive learning technologies, classified as

AI-mediated teaching practices, boast better learning outcomes, as research avers. However, they require a different skill set for the teacher and thus often changes the face of the job itself, which could scare the wits out of many. Above and beyond this, there is the question that students might soon lose a means of livability with the burgeoning artificial intelligence because that deprives them of the human correspondence necessary in developing analytical thinking and empathy towards pupils.

Changing Roles and Skills for Educators. AI in the classroom produces transformational changes to what exists in terms of the roles of an educator. One such case would characteristically be teacher to facilitator in learning and that is the direction in which critical thinking has changed to being done by teachers with pupils or students. In light of this transition, the demand on educators becomes one of possessing competencies to effectively use AI with their pedagogical technique. Such a professional development program, according to a report by the U.S. Department of Education, would be a boon for teachers because it prepares them to integrate AI in their instructional practice

Over-dependence on AI and the Necessity of Human Touch. AI technologies can carry out simple tasks and give versatile personalized learning, but students may not rely too much on any of those technologies. Such interaction empowers students to develop emotional intellect, social skills as well as critical thinking- things that cannot be attained through interacting with AI. Despite the benefits algorithmic systems may bring, models can reflect, inject, or exacerbate implicit and explicit societal biases into their outputs, disadvantaging certain demographic subgroups (Cabrera 2019).

Ethics and Bias Issues of Artificial Intelligence Systems. The AI system accepts some possible biases, which can be unfair or discriminatory to certain groups of students. The primary task now is to make AI equitable so as not to drive existing inequalities more deeply.

Ways to Deal with the Integration of AI. There are various modalities available to confront these challenges:

- Development of Professionals: Establishment of programs that train teachers to utilize the following amounts of learned AI tools in their teaching practice. This incorporates understanding the capabilities and limitations of AI while developing digital literacy skills.
- Complementing Integration: AI should complement rather than replace teacher instruction. One must keep the balance between technology with person-to-person interaction because the student must attain holistic development.
- Ethical Framework: The use of ethical frames in developing ethical frameworks prescribing uses of AI in education will address issues of data privacy, algorithmic bias, and the integrity of academic work.

The application of AI in education raises enormous challenges for pedagogy and thus requires careful consideration. Professional development, the balance between AI and human interaction, and ethical frameworks would support the establishment of an environment where the advantages of AI can be harnessed without sacrificing the critical human elements of teaching and learning.

2.5 Scalability and Resource Constraints

Really, bringing Artificial Intelligence (AI) into the education mosaic is full of promises but challenges galore, mainly scalability and resource constraints. AI technologies are inherently scalable; still, the initial stage of development and deployment requires many more resources in terms of finance and technology. This creates a risk in widening the gulf between educationally more deprivileged high-income and low-income regions, particularly in developing countries.

Financial Barriers on AI Implementation. All this is paralleled by the fact that quite a capital investment is needed in infrastructure, software, and people to make AI possible in education. Developing countries, in fact, have already budgetary constraints on such investments. Some of the aspects (mostly negative) associated with the implementation of AI in higher education are less generally addressed by the interviewed academics and are mainly related to the costs and efforts of implementing AI in higher education (Pisica 2023).

Technical Resource Constraints. Not only is there a financial pinch causing trouble on AI, but also very considerable technical infrastructure like reliable internet connectivity, advanced computing power, and well-trained personnel should be created. Among many of the developing regions of the earth, these essentials do not exist. The World Economic Forum notes the high price an advanced technology charges for developing countries constrained in their resources, making it a barrier to implementation.

The AI Divide: An Increasingly Worrisome Issue. In fact, this will be the horns of the growing "AI divide," which has come up due to the difference in adopting stage between developed and developing countries. It would intensify, in particular, educational inequality, as institutions in rich areas use AI for adjunct learning while the poorer ones have much less to supplement that. Brookings Institution asserts in this respect that infrastructure in Global South is inhibiting developing AI and thus depriving themselves from economic gains that follow such innovations.

Putting forward Solution and Strategy. These strategies can be employed to address such problems:

- International collaboration and funding: Developed countries and international agencies have a huge role to play in providing funds, sharing knowledge, and making AI technology accessible to educational institutions in developing countries.
- Open-Source AI Tools: Open-source AI platforms are going to bring down the price and facilitate the use of AI. Adapting the software to meet internal educational purposes should be relatively easy as opposed to paying a lot for proprietary software.

- Capacity Building: Training should be designed to create a local cadre that possesses specialized knowledge for the purpose of positioning educators and administrators to implement and manage AI systems.
- Policy Development: There needs to be a concrete adoption of policies to encourage AI use in education by governments in developing countries, such as incentives to motivate the private sector's participation as well as establishment of a regulatory framework for equitable access.

Therefore, it is of utmost importance since it brings education revolutionized but has serious resource constraints in developing countries, hence unable to scale. The effort should be made concertedly by the entire global populace to see to it that the advantages of AI in education are extended to all.

3 Ethical Considerations in AI Integration

Integrating AI into Industry 4.0 and sustainability frameworks necessitate addressing ethical challenges for responsible development. Key among these are accountability and transparency, which are essential for fostering trust in AI systems, particularly in educational contexts where automated decision-making can significantly impact student outcomes (Conijn 2023, & Bai 2024). Furthermore, the issues of intellectual property and content ownership arise as AI systems generate content, necessitating clear guidelines to protect creators' rights (Langman 2021 & Adelakun 2024). The ethics of automation also come into play, as the reliance on AI can lead to job displacement and erosion of critical skills, highlighting the need for a balanced approach that promotes human autonomy (Strauß, 2021 & BaHammam 2023). Additionally, cultural sensitivity is crucial, as AI applications developed in one context may not be appropriate in another, requiring adaptations to local norms (Salih 2024 & Chatzivasileiou 2024). Lastly, to reduce the hazards associated with over-reliance on AI technology in educational settings, it is vital to promote student autonomy and critical thinking (Bai, 2024 & Miao, 2024).

3.1 Accountability & Transparency

Opaque decision-making processes within artificial intelligence (AI) systems—often referred to as "black boxes" present serious ethical challenges, especially in sensitive fields like healthcare and education. These "black boxes" obscure how AI arrives at its conclusions, raising critical questions about accountability and the risk of embedded biases. Without transparency, stakeholders cannot understand how decisions are made, which can lead to mistrust and inequitable outcomes (Sqalli 2023 & Ueda 2023).

Explainable AI (XAI) aims to mitigate these issues by providing insights into AI decision-making processes, thereby enhancing accountability and fostering trust among users (Familoni, 2024 & Qadir, 2022). Furthermore, ethical frameworks that emphasize fairness and transparency are essential for preventing biases in AI systems, ensuring that these technologies promote equitable outcomes for all users (Obasa 2023). The integration of AI in various industries necessitates a commitment to sustainability, where ethical considerations must align with technological advancements to support responsible practices (Odeyemi, 2024). By prioritizing accountability and transparency, organizations can not only comply with regulatory standards but also cultivate a culture of ethical AI that addresses societal concerns and enhances stakeholder confidence (Aldoseri 2023). Ultimately, the responsible application of AI, characterized by clear accountability and transparent processes, is vital for preventing biases, promoting trust, and ensuring equitable outcomes across diverse applications (Richardson, 2021 & Wang, 2023).

3.2 Intellectual Property and Content Ownership

The integration of AI within Industry 4.0 presents significant potential for enhancing operational efficiency and fostering environmental sustainability, while also necessitating careful consideration of ethical implications. As AI systems, such as chatbots and virtual tutors, generate content, questions arise regarding who owns this content. Current legal frameworks struggle to address the ownership of AI-generated works, leading to potential copyright infringements and disputes over intellectual property rights (Lima 2020). Since the authorship and ownership of AI-generated work are ambiguous, certain rules and regulations are required. These need to guarantee that authors and developers are properly credited and given the credit they deserve (Wittenberg 2024). Additionally, ethical issues surrounding AI in education -specifically, individualized learning - emphasize the necessity of responsible AI activities. Innovation promotion and intellectual property rights protection must be balanced in these activities (Tonbuloglu, 2023 & Hashim, 2022).

Ethical considerations are of paramount importance in the context of AI integration within Industry 4.0, as AI offers significant potential for enhancing operational efficiency and fostering environmental sustainability (Valeriya 2024 & Cioffi 2020). However, without robust policies, the risk of misuse and ethical violations remains high. Establishing trust and responsibility in AI applications thus need extensive frameworks that include content ownership, precise citation, and ethical usage (Oyewole, 2024). Such frameworks not only protect intellectual property but also encourage sustainable practices in AI deployment, ultimately contributing to a more ethical and responsible integration of AI technologies across various domains (Vinchon, 2023 & Wehlmann, 2024).

3.3 Ethics of Automation

The ethical ramifications of automation must be carefully considered when integrating artificial intelligence (AI) across many industries, especially with relation to worker dynamics, administrative procedures, and environmental sustainability. As AI systems increasingly automate routine administrative functions, concerns arise regarding job displacement and the social roles of workers. In contrast to its potential to boost productivity and create new jobs, AI also carries a significant risk of displacing workers, with low-skilled workers being particularly vulnerable (Georgieff, 2022, Liang, 2024 & Acemoğlu, 2018). Therefore, achieving a proper balance in AI integration is essential to ensure that job security is not compromised. This balance can be facilitated through robust regulatory frameworks that promote ethical AI practices while safeguarding workers' rights (Jarota, 2021 & Toli, 2024).

Industry 4.0 AI integration necessitates an emphasis on sustainable automation, striking a balance between financial benefits and social and environmental effects (Bughin, 2023 & Kumar, 2023). In order to use AI ethically, possible problems such as growing wealth disparities must be addressed, and people impacted by automation must be treated fairly (Liang, 2024 & Wang, 2023). To reduce risks and build stakeholder confidence, AI decision-making must be transparent (Felzmann 2020 & Yan 2023). Successful AI integration ultimately depends on moral standards that put worker welfare and productivity first, opening the door to a sustainable workplace in the era of automation (Moradi 2020 & Bankins 2023).

3.4 Cultural Sensitivity and Context

The successful and ethical integration of AI into diverse socio-economic environments depends on a strong understanding of cultural nuances and contextual factors. The cultural sensitivity required for effective worldwide deployment is sometimes absent from AI systems based on Western paradigms. This oversight can lead to misalignment with local cultural norms and practices, particularly in regions where socio-economic conditions differ significantly from those of the developers (Salih, 2024). For instance, language models trained primarily on English texts may inadvertently perpetuate biases, failing to account for the linguistic and cultural diversity present in non-English-speaking populations (Sato 2024). Cultural sensitivity in the design of AI systems is essential for ensuring their adaptability and relevance in various contexts (Balasubramanian 2023).

Sustainable approaches to cultural sensitivity are essential for successful AI integration in Industry 4.0. This involves not only the ethical design of AI systems but also the active engagement of local communities in the development process, thereby empowering marginalized voices and perspectives (Karpouzis, 2024). Human-centered design, according to Shneiderman (2022), enables AI to more accurately reflect the many cultural identities and values of users, increasing its efficiency and acceptability. As AI continues to evolve, it is imperative that developers prioritize cultural context and sensitivity to mitigate biases and foster equitable outcomes across different socio-economic settings (Osasona 2024 & Yuan 2024).

3.5 Student Autonomy and Critical Thinking

The benefits and downsides of integrating artificial intelligence (AI) into education are complicated, especially when it comes to student autonomy and critical thinking. Although AI presents the possibility of individualized educational experiences that encourage self-reliance and creativity, enabling learners to interact with content in ways that suit their unique learning preferences (Grubaugh 2024), it also presents issues. However, there is a growing concern regarding the potential for over-reliance on AI, which may lead to shallow learning experiences where students become passive consumers of information rather than active participants in their education (Chan, 2023). In order to mitigate this potential negative consequence, it is imperative that AI systems are conceived as supportive of, rather than substitute for, human cognition, functioning as tools that promote critical thinking and problem-solving capacities (Boscardin, 2023).

Promoting student autonomy and critical thinking in the context of AI and Industry Revolution 4.0 (IR 4.0) requires a strong emphasis on sustainability. Teachers should help students become more AI literate so they can assess AI-generated information critically and make wise decisions about using it (Du 2023). Effective AI integration in education requires a careful balance: leveraging AI's potential while simultaneously nurturing essential human skills such as critical thinking and independent thought. This balanced approach not only prepares students for an increasingly AI-centric world but also promotes a deeper understanding of the ethical considerations surrounding this technology (Moustaghfir 2024).

4 How Challenges and Ethical Issues Can Be Addressed

Artificial intelligence (AI) technology has the potential to completely change several facets of our lives, including healthcare, banking, entertainment, transportation, and education. However, as AI continues to grow, concerns about privacy and data security have gained traction. Organizations may successfully overcome obstacles and protect privacy in AI by addressing important risks, implementing strong governance and security measures, and embracing privacy-enhancing activities.

4.1 Policy Frameworks and Governance

To guarantee that AI in higher education is applied responsibly, ethically, and inclusively, a strong legislative framework and governance mechanism must be established. In addition to encouraging creativity and inclusion, these frameworks ought to include rules and policies pertaining to data privacy, algorithmic justice, and accountability.

Data Privacy Protection: Institutions need to make sure that learners and teachers understand exactly how their data will be gathered, saved, and utilized. The opt-in and opt-out procedures must be transparent. Only authorized individuals and systems should have access to sensitive data, and appropriate audit trails should be in place to monitor data usage. Policies must comply with national and international data protection regulations, such as the Data Protection Act of India, the General Data Protection Regulation (GDPR), and Health Insurance Portability and Accountability Act (HIPAA) (Chan, 2023). Data Protection Officers (DPOs) may be appointed to oversee compliance and ensure continuous evaluation of privacy safeguards.

Algorithmic Fairness: Assess AI algorithms on a regular basis to find and reduce biases in results. AI systems ought to be made to cater to a wide range of demographics, including those with different socioeconomic, linguistic, and cultural backgrounds. To foster trust, institutions must provide stakeholders with access to the standards and reasoning underlying AI decision-making processes. To ensure equity in the application of AI, Ethics Committees are to be formed with representatives from the community, educators, and AI specialists.

Accountability Mechanisms: Establishing who is accountable for AI-driven decision developers, administrators, or the institutions themselves is a necessary step. The staff and students must be provided with clear procedures for reporting and resolving problems brought on by AI choices (Wong 2024). AI systems ought to keep thorough records of their decisions so that results may be linked to particular inputs and guidelines. Create liability frameworks that specify how users, institutions, and developers share accountability.

Encouraging Innovation and Inclusivity: Create regulatory sandboxes so that AI tools can be evaluated in safe settings without worrying about breaking any laws. Provide funding, tax breaks, or other forms of assistance to create cutting-edge AI systems specifically suited for the educational sector. By encouraging free or subsidized access, policies should guarantee that AI technologies are available to everyone, irrespective of socioeconomic background (Jin 2024). To lower barriers for underrepresented communities, promote the creation of AI systems that are sensitive to regional languages and cultural situations.

Collaborative Governance: Regional or national AI policies should be overseen and created. Create institution-specific policies that meet local demands and are in line with more general regulations. Moral AI systems should be jointly developed with tech firms. Faculty and students should actively include end users in the development of policies to guarantee their applicability and acceptance (Livingstone 2024).

4.2 Capacity Building and Training

To mitigate the challenges and ethical problems and promote an inclusive and moral AI environment, capacity building and training are essential.

Increasing Faculty and Student Knowledge of AI: Many students and educators lack the fundamental knowledge needed to use AI technologies effectively. Plan certification programs, workshops, and seminars on the principles, resources, and uses of AI. Introduce classes on AI in a variety of subject areas to improve interdisciplinary comprehension. For upskilling initiatives, create alliances with AI companies or online learning environments.

Ethical AI Education and Awareness Challenge: Inequality, unjust grading, and privacy invasion are just a few of the ethical conundrums that can arise from the misuse of AI and algorithmic bias (Browne 2023). Include courses on AI ethics in faculty professional development programs and student curricula. In training sessions, promote conversation on practical ethical issues with AI. Encourage awareness of ethical AI practices, such as responsibility, transparency, and justice.

Capacity Building in Data Handling and Privacy Protection Challenge: Managing sensitive student and institutional data raises privacy concerns. Train stakeholders in data governance policies and compliance with regulations like General Data Protection Regulation (GDPR) or Family Educational Rights and Privacy Act (FERPA). Offer specialized training on secure data handling and anonymization techniques. Raise awareness about data misuse and its consequences.

Overcoming Opposition to AI Adoption Challenge: Because they don't trust the technology or are afraid of losing their jobs, faculty and administrators may be reluctant to integrate AI. Provide training that focuses on how AI enhances human jobs rather than takes their place. Give practical examples of how AI can be used to improve administrative, instructional, and learning activities. Engage educators in the creation and application of AI tools to promote acceptance and ownership.

Technical Training and Support Infrastructure Challenge: Insufficient technical know-how may make it more difficult to apply AI effectively (Brandon 2023). To set up AI resource centers manned by qualified staff to assist instructors and students. Continued instruction in AI platforms and technologies pertinent to administrative and academic requirements should be given. Provide access to AI infrastructure, such as cutting-edge software and cloud-based platforms.

Promoting Inclusion and Diversity: Some groups could not have access to AI training opportunities, and AI tools have the potential to reinforce prejudices. Create inclusive training initiatives that guarantee underrepresented groups are represented in AI-related disciplines. Encourage the use of a variety of datasets and make sure that biases in AI algorithms are taken into account during training. Encourage programs that promote minority and female participation in AI education.

Ongoing Professional Development Challenge: AI is a quickly changing discipline that necessitates constant adjustment. Offer faculty members options for ongoing education through sabbaticals, higher certificates, and short courses. To stay current on trends and best practices, encourage attendance at AI conferences, webinars, and international forums.

Development and Application of Ethical Policies: The absence of explicit ethical standards may lead to misunderstandings and improper usage of AI in the classroom. Educate administrators and professors on how to create and implement moral AI policies in their organizations (Moore 2024). Form committees to supervise the moral application of AI and handle any problems that may arise.

4.3 Interdisciplinary Collaboration

Interdisciplinary Collaboration plays a pivotal role in addressing challenges and ethical issues in integrating AI into higher education. By fostering cooperation among diverse fields, institutions can ensure AI's responsible and effective use while mitigating risks.

Development and Application of Ethical AI: The challenge is to guarantee accountability, transparency, and equity in AI systems. To develop and evaluate AI systems, assemble multidisciplinary teams comprising computer scientists, ethicists, sociologists, and educators. Work together to establish ethical AI development frameworks that take psychological, cultural, and societal aspects into account. From alliances with law and humanities departments to including morality into AI applications.

Reducing Algorithmic Bias Challenge: Unfair results may result from AI systems reflecting biases in data or design. To find and reduce biases in datasets and algorithms, consult statisticians, social scientists, and subject matter experts. Create procedures for assessing AI tools from a variety of discipline perspectives while maintaining inclusivity and equity. Encourage teamwork in research to improve data gathering procedures and create techniques for detecting bias.

Improving Data Security and Privacy Challenge: Preventing unauthorized use of private student and institutional information. To build strong data governance frameworks, integrate knowledge from computer science (cybersecurity), law (data legislation), and ethics. Collaborate with behavioural scientists to comprehend user concerns and create privacy-focused AI products (Gerdes, 2022). Create multi-disciplinary task forces to monitor adherence to privacy laws such as FERPA or GDPR.

Creating AI Solutions Focused on Students: The challenge is to make sure AI tools meet the various needs of instructors and students. Engage psychologists, educators, and AI engineers in the creation of technologies that accommodate various learning requirements and styles. Work together with curriculum designers to incorporate AI into instruction in ways that improve student learning. Assess how AI affects student motivation, engagement, and wellbeing by using knowledge from the social sciences.

Promoting the Development of Inclusive AI: Underrepresented groups might not be able to influence the creation and application of AI. To guarantee inclusivity, incorporate viewpoints from disability advocates, gender studies, and cultural studies. Create AI systems that, with interdisciplinary input, take into consideration the various demands of underrepresented groups. Encourage partnerships that concentrate on regional and global differences in AI impact and accessibility.

4.4 Ethical AI Design Principles

A multi-stakeholder approach involving cooperation between educators, legislators, researchers, developers, and other players in the AI ecosystem is necessary to address the ethical concerns of AI in education. Among the potential courses of action are:

Creating ethical frameworks and guidelines: These can offer a shared vision as well as a set of values and principles to direct the moral development, application, deployment, and impact of artificial intelligence in education. For instance, in November 2021, UNESCO released the "Recommendation on the Ethics of Artificial Intelligence," the first-ever worldwide guideline on AI ethics.

Applying ethical standards and practices: To guarantee the moral performance and conduct of AI systems in education, ethical standards, and practices can operationalize ethical frameworks and guidelines into specific acts and procedures (Pistilli 2023). Data quality assurance, algorithmic transparency and explainability, human oversight and intervention, privacy and data protection rules, impact assessment and evaluation, stakeholder participation and consultation, and so on are a few examples of ethical practices and standards.

Encouraging ethical education and awareness: Students, instructors, and other stakeholders in the AI ecosystem can become more ethically literate and competent through ethical education and awareness (Luo 2024). Curriculum integration, professional development, public involvement, media literacy, digital citizenship, and so forth are a few examples of ethical education and awareness.

Strengthening ethical governance and regulation: These practices can offer an institutional and legal framework to guarantee the responsibility and accountability of AI actors in education. Laws and rules, codes of conduct, ethics committees, oversight organizations, complaint and redress procedures, and so forth are examples of ethical governance and regulation.

4.5 Community Engagement

One effective strategy for tackling the difficulties and moral dilemmas associated with incorporating AI into higher education is community participation. Institutions may guarantee inclusivity, equity, and transparency by actively including the larger community - students, instructors, parents, neighbourhood organizations, and industry stakeholders - in the planning, creation, and execution of AI projects.

Encouraging Inclusivity and Equity Difficulty: AI systems may unintentionally leave out marginalized communities or maintain existing disparities. Work together with local communities to comprehend their demands and make sure AI solutions fairly answer them. Collaborate with advocacy organizations and NGOs to provide underprivileged communities with AI resources and education.Create AI systems with feedback from all facets of the community to guarantee social and cultural inclusion.

Reducing Concerns About AI-Related Job Losses: Staff and educators are especially worried about AI taking the place of humans. Talk to the community about how AI enhances human roles rather than taking their place. Give staff and academic members opportunity for upskilling and retraining. Emphasize instances when AI has improved job happiness and productivity.

Promoting the Use of AI Responsibly: Ethical and legal problems may arise if students or other parties misuse AI. Run awareness campaigns about the possible hazards of using AI tools responsibly. Create community-driven guidelines for using AI in classroom (Bousquette 2024). Involve legal professionals and law enforcement in developing protections against AI abuse.

Developing Collaborations for Ethical AI Development Challenge: Insufficient funds or knowledge to handle intricate ethical AI problems. Work together to jointly build ethical AI solutions with regional tech firms, think tanks, and educational institutions. Funding and assisting community-driven AI initiatives should involve both the public and business sectors (Eden 2024). Create mentorship programs where teachers and students receive guidance on appropriate AI techniques from industry experts.

Challenge of Promoting a Collaborative Research Ecosystem: Research initiatives that are isolated from one another impede creativity and community engagement. Collaborate on AI research and projects with nearby schools, libraries, and community colleges. Engage community members and students as co-researchers in AI projects. Organize public research showcases to disseminate findings and promote wider involvement.

References

Acemoğlu, D., & Restrepo, P. (2018). Artificial intelligence, automation, and work, The Economics of Artificial Intelligence: *An Agenda*, doi: https://doi.org/10.3386/w24196

Adelakun, B., Majekodunmi, T., & Akintoye, O. (2024). AI and ethical accounting: Navigating challenges and opportunities. *The International Journal of Advances in Management and Economics, 6*(6), 224–241. https://doi.org/10.51594/ijae.v6i6.1230

Agarwal, R., Bjarnadottir, M., Rhue, L., Dugas, M., Crowley, K., Clark, J., & Gao, G. (2023). Addressing algorithmic bias and the perpetuation of health inequities: An AI bias aware framework. *Health Policy Technol, 12*, 100702. https://doi.org/10.1016/j.hlpt.2022.100702

Aldoseri, A., & Al-Khalifa, K. (2023). Re-thinking data strategy and integration for artificial intelligence: concepts, opportunities, and challenges. *Applied Sciences, 13*(12), 7082. https://doi.org/10.3390/app13127082

BaHammam, A. (2023). Balancing innovation and integrity: the role of AI in research and scientific writing. *Nature and Science of Sleep, 15*, 1153–1156. https://doi.org/10.2147/nss.s455765

Bai, X. (2024). The role and challenges of artificial intelligence in information technology education. *Pacific International Journal, 7*(1), 86–92. https://doi.org/10.55014/pij.v7i1.524

Baker, R. S., & Hawn, A. (2022). Algorithmic bias in education. *International Journal of Artificial Intelligence in Education, 1*, 1–41. https://doi.org/10.1007/s40593-021-00285-9

Balasubramanian, S., Devarajan, H. R., Raparthi, M., Dodda, S. B., Maruthi, S., & Adnyana, M. D. M. (2023). Ethical considerations in AI-assisted decision-making for end-of-life care in healthcare. *Pacific Time Zone, 47*(4), 167–182. https://doi.org/10.52783/pst.168

Bankins, S., Ocampo, A. C., Marrone, M., Restubog, S. L. D., & Woo, S. E. (2023). A multilevel review of artificial intelligence in organizations: implications for organizational behavior research and practice. *Journal of Organizational Behavior Research, 45*(2), 159–182. https://doi.org/10.1002/job.2735

Boscardin, C. K., Gin, B., Golde, P. B., & Hauer, K. E. (2023). Chatgpt and generative artificial intelligence for medical education: potential impact and opportunity. *Academic Medicine, 99*(1), 22–27. https://doi.org/10.1097/acm.0000000000005439

Bousquette, I. 2022. Inside Universities Love-Hate Relationship with ChatGPT. https://www.wsj.com

Brandon, E., Eaton, L., Gavin, D., & Papini, A.: Cross-Campus Approaches to Building a Generative AI Policy. (2023)

Browne, J., Drage, E., & Mc Inerney, K: Tech Workers Perspectives on Ethical Issues in AI Development. (2023). Big Data & Society, 11(1), 205–220. doi: https://doi.org/10.1177/20539517231221780

Bughin, J. (2023). Does artificial intelligence kill employment growth: the missing link of corporate AI posture. *Frontiers in Artificial Intelligence, 6*. https://doi.org/10.3389/frai.2023.1239466

Cabrera, Á. A., Epperson, W., Hohman, F., Kahng, M., Morgenstern, J., & Chau, D. H. (2019). *FAIRVIS: Visual analytics for discovering intersectional bias in machine learning*. In: IEEE Conference Visual Analytical Science & Technology. (VAST), pp. 46–56. doi: https://doi.org/10.1109/VAST47406.2019.8986948

Chan, C. K. Y. (2023). A comprehensive AI policy education framework for university teaching and learning. *International Journal of Educational Technology in Higher Education, 20*(1), 20–35. https://doi.org/10.1186/s41239-023-00408-3

Chan, C. K. Y., & Hu, W. (2023). Students voices on generative AI: perceptions, benefits, and challenges in higher education. *International Journal of Educational Technology in Higher Education, 20*, 43. https://doi.org/10.1186/s41239-023-00411-8

Chatzivasileiou, D. (2024). AI, international relations & religion. *Journal of Politics and Ethics in New Technologies and AI, 3*(1), e37109. https://doi.org/10.12681/jpentai.37109

Cioffi, R., Travaglioni, M., Piscitelli, G., Petrillo, A., & Felice, F. (2020). Artificial intelligence and machine learning applications in smart production: progress, trends, and directions. *Sustainability, 12*(2), 492. https://doi.org/10.3390/su12020492

Conijn, R., Kahr, P., & Snijders, C. (2023). The effects of explanations in automated essay scoring systems on student trust and motivation. *International Journal of Learning Analytics and Artificial Intelligence in Education, 10*, 37–53. https://doi.org/10.18608/jla.2023.7801

Du, X., & Wang Xi (2023). Play by design: developing artificial intelligence literacy through game-based learning. *Journal of Computer Science Research. 5*(4), 1–12. doi:https://doi.org/10.30564/jcsr.v5i4.5999

Eden, C. A., Chisom, O. N., & Adeniyi, I. S. (2024). Integrating AI in education: Opportunities, challenges, and ethical considerations. *Magna Scientia Advanced Research and Reviews, 10*, 006–013. https://doi.org/10.30574/msarr.2024.10.2.0039

Familoni, B. (2024). Ethical frameworks for AI in healthcare entrepreneurship: a theoretical examination of challenges and approaches. *International Journal of Frontiers in Biology and Pharmacy Research, 5*(1), 057–065. https://doi.org/10.53294/ijfbpr.2024.5.1.0032

Felzmann, H., Fosch-Villaronga, E., Lutz, C., & Tamò-Larrieux, A. (2020). Towards transparency by design for artificial intelligence. *Science and Engineering Ethics, 26*(6), 3333–3361. https://doi.org/10.1007/s11948-020-00276-4

Georgieff, A., & Hyee, R. (2022). Artificial intelligence and employment: new cross-country evidence. *Frontiers in Artificial Intelligence, 5*. https://doi.org/10.3389/frai.2022.832736

Gerdes, A. (2022). The tech industry hijacking of the AI ethics research agenda and why we should reclaim it. *Discover Artificial Intelligence, 2*, 25. https://doi.org/10.1007/s44163-022-00043-3

Grubaugh, S., & Levitt, G. (2024). The future of elementary social studies: Harnessing AI's potential through evidence-based practices. *Technium Social Sciences Journal, 58*, 87–93. https://doi.org/10.47577/tssj.v58i1.10991]

Hashim, S., Omar, M., Jalil, H., & Sharef, N. (2022). Trends on technologies and artificial intelligence in education for personalized learning: systematic literature review. *International Journal of Academic Research in Progressive Education and Development, 12*(1), 884–903. https://doi.org/10.6007/ijarped/v11-i1/12230

Jarota, M. (2021). Artificial intelligence and robotisation in the EU - should we change OHS law. *Journal of Occupational Medicine and Toxicology, 16*(1). https://doi.org/10.1186/s12995-021-00301-7

Jin, Y., Yan, L., Echeverria, V., & Gašević, D., Martinez-Maldonado, R.: *Generative AI in higher education: A global perspective of institutional adoption policies and guidelines*. https://arxiv.org/abs/2405.11800

Karpouzis, K. (2024). Platos shadows in the digital cave: controlling cultural bias in generative AI. *Electronics, 13*(8), 1457. https://doi.org/10.3390/electronics13081457

Kumar, A., Bhattacharyya, S. S., & Krishnamoorthy, B. (2023). Automation-augmentation paradox in organizational artificial intelligence technology deployment capabilities; an empirical investigation for achieving simultaneous economic and social benefits. *Journal of Enterprise Information Management, 36*(6), 1556–1582. https://doi.org/10.1108/jeim-09-2022-0307

Langman, S., Capicotto, N., Maddahi, Y., & Zareinia, K. (2021). Roboethics principles and policies in Europe and North America, SN. *Applied Sciences, 3*(12). https://doi.org/10.1007/s42452-021-04853-5

Liang, Y. (2024). The impact of artificial intelligence on employment and income distribution. *Journal of Education, Humanities and Social Sciences, 27*, 166–171. https://doi.org/10.54097/2a7a8830

Lima, G., Kim, C., Ryu, S., Jeon, C., & Cha, M. (2020). Collecting the Public Perception of AI and Robot Rights. *Proceedings ACM Hum. Comput. Interact, 4*, 1–24. https://doi.org/10.1145/3415206

Livingstone, V. *I quit teaching because of ChaGPT*. https://time.com

Luo Jess, J. (2024). A critical review of GenAI Policies in higher education assessment: a call to reconsider the originality of students work. *Assessment & Evaluation in Higher Education, 49*(5), 651–664. https://doi.org/10.1080/02602938.2024.2309963

Miao, J., Thongprayoon, C., Suppadungsuk, S., Garcia, V. O. A., Qureshi, F., & Cheungpasitporn, W. (2024). Ethical dilemmas in using AI for academic writing and an example framework for peer review in nephrology academia: a narrative review. *Clinics and Practice, 14*(1), 89–105. https://doi.org/10.3390/clinpract14010008

Moore, S., & Lookadoo, K. (2024). Communicating Clear Guidance: Advice for Generative AI Policy Development in Higher Education. *Business and Professional Communication Quarterly, 87*(4), 610–629. https://doi.org/10.1177/23294906241254786

Moradi, P., & Levy, K. (2020). *The future of work in the age of AI, 269–288*. https://doi.org/10.1093/oxfordhb/9780190067397.013.17

Moustaghfir, S., & Brigui, H. (2024). Navigating critical thinking in the digital era: an informative exploration. *International Journal of Linguistics, Literature and Translation, 1*, 137–143m. https://doi.org/10.32996/ijllt.2024.7.1.11x

Obasa, A., Palk, A. Responsible application of artificial intelligence in healthcare, *South African Journal of Science. 119*(5/6). doi: https://doi.org/10.17159/sajs.2023/14889

Odeyemi, O. (2024). Reviewing the role of AI in fraud detection and prevention in financial services. *International Journal of Science and Research Archive, 11*(1), 2101–2110. https://doi.org/10.30574/ijsra.2024.11.1.0279

Osasona, F., Daraojimba, A. I., Atadoga, A., Onwusinkwue, S., Obi, O. C., & Dawodu, S. O. (2024). AI integration in business analytics: a review of USA and African trends. *Computer Science & IT Research Journal, 5*(2), 432–446. https://doi.org/10.51594/csitrj.v5i2.793

Oyewole, A., Adeoye, O. B., Addy, W. A., Okoye, C. C., Ofodile, O. C., & Ugochukwu, C. E. (2024). Promoting sustainability in finance with AI: a review of current practices and future potential. *World Journal of Advanced Research and Reviews: International, 21*(3), 590–607. https://doi.org/10.30574/wjarr.2024.21.3.0691

Pisica, A. I., Edu, T., Zaharia, R. M., & Zaharia, R. (2023). Implementing artificial intelligence in higher education: Pros and cons from the perspectives of academics. *Societies, 13*, 118. https://doi.org/10.3390/soc13050118

Pistilli, G., Muñoz Ferrandis, C., & Jernite, Y., Mitchell, M. Stronger together: On the articulation of ethical charters, legal tools, and technical documentation in ML. *Proceedings of the 2023 ACM Conference on Fairness, Accountability, and Transparency* (pp. 343–354). https://arxiv.org/abs/2305.18615

Qadir, J., Islam, M., & Al-Fuqaha, A. (2022). Toward accountable human-centered AI: rationale and promising directions. *Journal of Information, Communication and Ethics in Society, 20*(2), 329–342. https://doi.org/10.1108/jices-06-2021-0059

Richardson, J., Smith, C., Curtis, S., Watson, S., Zhu, X., Barry, B., & Sharp, R. R. (2021). Patient apprehensions about the use of artificial intelligence in healthcare. *NPJ Digital. Medicine, 4*(1), 140. https://doi.org/10.1038/s41746-021-00509-1

Saaida, M. B. (2023). AI-driven transformations in higher education: Opportunities and challenges. *International Journal of Educational Research and Studies, 5*, 29–36. https://doi.org/10.5281/zenodo.8164414

Salih, S. (2024). Perceptions of faculty and students about use of artificial intelligence in medical education: a qualitative study. *Cureus, 16*(4)). https://doi.org/10.7759/cureus.57605

Sato, K., Kaneko, H., & Fujimura, M. (2024). *Reducing cultural hallucination in non-English languages via prompt engineering for large language models.* https://doi.org/10.31219/osf.io/4hzya

Shneiderman, B.. (2022). Human-centered AI, Retrieved 19 Dec. 2024 from. doi:https://doi.org/10.1093/oso/9780192845290.001.0001

Slimi, Z., Carballido, B. V.: Navigating the ethical challenges of artificial intelligence in higher education: An analysis of seven global AI ethics policies. TEM J. 12, (2) (2023). doi: https://doi.org/10.18421/TEM122-02

Sqalli, M., Begali, A., Mukhammadjon, G., & Shokhrukhbek, N. (2023). Humanizing AI in medical training: ethical framework for responsible design. *Frontiers in Artificial Intelligence, 6*. https://doi.org/10.3389/frai.2023.1189914

Strauß, S. (2021). Deep automation bias: how to tackle a wicked problem of AI? *Big Data Cognitive. Computing, 5*(2), 18. https://doi.org/10.3390/bdcc5020018

Toli, L., & Manasa, G. M. (2024). Artificial intelligence: opportunities and challenges for the social education and profession. *Scholars Bull., 10*(04), 143–147. https://doi.org/10.36348/sb.2024.v10i04.005

Tonbuloglu, B. (2023). An evaluation of the use of artificial intelligence applications in online education. *Journal of Educational Technology and Online Learning, 6*(4), 866–884. https://doi.org/10.31681/jetol.1335906

Ueda, D., Kakinuma, T., Fujita, S., Kamagata, K., Fushimi, Y., Ito, R., Matsui, Y., Nozaki, T., Nakaura, T., Fujima, N., Tatsugami, F., Yanagawa, M., Hirata, K., Yamada, A., Tsuboyama, T., Kawamura, M., Fujioka, T., & Naganawa, S. (2023). Fairness of artificial intelligence in healthcare: review and recommendations. *Japanese Journal of Radiology, 42*(1), 3–15. https://doi.org/10.1007/s11604-023-01474-3

Valeriya, G., Kirola, M., Gupta, M., Bharathi, P., Acharya, P., & Sharma, R. (2024). Evaluating the impact of AI-based sustainability measures in industry 5.0: a longitudinal study. *Bio Web of Conferences., 86*, 01058. https://doi.org/10.1051/bioconf/20248601058

Vinchon, F., Lubart, T., Bartolotta, S., Gironnay, V., Botella, M., Bourgeois-Bougrine, S., Burkhardt, J.-M., Bonnardel, N., Corazza, G. E., Glăveanu, V., Hanchett Hanson, M., Ivcevic, Z., Karwowski, M., Kaufman, J. C., Okada, T., Reiter-Palmon, R., & Gaggioli, A. (2023). Artificial intelligence & creativity: a manifesto for collaboration. *The Journal of Creative Behavior, 57*(4), 472–484. https://doi.org/10.1002/jocb.597

Wang, C., Liu, S., Yang, H., Jiu-lin, G., Wu, Y., & Liu, J. (2023). Ethical considerations of using ChatGPT in healthcare. *Journal of Medical Internet Research, 25*, e48009. https://doi.org/10.2196/48009

Wang, X. (2023). Research on the realization path of inclusive employment structure under the application of artificial intelligence. *Study of social and economic policy, 3*(2), 124–127. https://doi.org/10.26480/seps.02.2023.124.127

Wehlmann, C. (2024). Resilient and sustainable AI. Positioning paper on the relation of AI, *Resilience and Sustainability*, 5–19. doi: https://doi.org/10.1007/978-3-658-43705-3_2

Wittenberg, C., Epstein, Z., Berinsky, A.J., Rand, D. G. (2024). Labeling AI-generated content: promises, perils, and future directions, doi:https://doi.org/10.21428/e4baedd9.0319e3a6

Wong, M. Another disastrous year inf ChatGPT school is beginning. https://www.theatlantic.com

Yan, L., Sha, L., Zhao, L., Li, Y., Martinez-Maldonado, R., Chen, G., Li, X., Jin, Y., & Gašević, D. (2023). Practical and ethical challenges of large language models in education: a systematic scoping review. *British Journal of Educational Technology, 55*(1), 90–112. https://doi.org/10.1111/bjet.13370

Yuan, X., Hu, J., & Zhang, Q. (2024). *A comparative analysis of cultural alignment in large language models in bilingual contexts*. https://doi.org/10.31219/osf.io/6hpcf

Future Directions and Policy Implications: AI in Higher Education

Osamah Mohammed Jasim, Pushkar Baranwal, Nitisha Aggarwal, Abhirup Nandy, Geetika Jain Saxena, Amit Pundir, and Sanjeev Singh

Abstract The involvement of AI in higher education requires solid policies that guide its use and help ensure it is both ethical and effective. This chapter reviews the range of challenges and benefits linked to using AI in education, calling for laws that are capable of keeping up with fast changes in technology. By examining current government rules in detail, the chapter points out serious deficiencies in present laws and provides complete methods for solving legal, social, and technical challenges. The topic addresses models for working with stakeholders, ways to cooperate with other countries, AI literacy programmes, auditing of the system, and what to invest in. Integrating insight from worldwide examples and multiple academic fields, this chapter proposes a unified guideline for policy creation that keeps in mind both innovations and responsibility, equity, as well as efficiency and technological progress, while respecting human needs. According to the framework, adaptable structures in governance should adapt with technology, yet keep ethical standards. At the end of the chapter, steps are suggested for policymakers, those leading institutions, and teachers, highlighting that using AI in higher education works best if the rules and culture support ethics, democracy, and fairness in the digital age.

Keywords Artificial intelligence · Higher education · Sustainable development goals · Inclusive education · Resource management · Ethical AI

1 Introduction

Rapidly increasing use of artificial intelligence in all areas of higher learning has prompted major changes in how people teach, manage education, conduct research, and interact with students. Because these advances open up exciting possibilities for

---

O. M. Jasim · P. Baranwal · N. Aggarwal · A. Nandy · S. Singh
Institute of Informatics and Communication, South Campus, University of Delhi, Delhi, India

G. J. Saxena · A. Pundir (✉)
Department of Electronics, Maharaja Agrasen College, University of Delhi, Delhi, India
e-mail: amitpundir@mac.du.ac.in

A. Ara (ed.), *Role of Artificial Intelligence in Higher Education*,
https://doi.org/10.1007/978-981-95-5350-1_6

innovation and boost efficiency, they also make it difficult to solve the challenges they introduce. Since AI technology is becoming more common in education, there is now a strong need to put in place governance rules that can manage its various social, ethical, and technological impacts (Holmes & Porayska Pomsta, 2022). Many traditional policies fail to adequately address the fast changes happening in AI. Since both technological change and machine learning systems are developing quickly, there is a need for a new way of creating and applying educational policies (Zawacki-Richter et al., 2019). The chapter calls for flexible policies and systems that can grow with technology, all the while sticking firmly to key ethical rules and values for education. This policy situation is more significant than it may first appear. Failing to think carefully about how AI is put into use in education may cause it to widen current gaps, infringe on people's privacy and independence, give rise to biased decisions, and undermine education as we know it (Reich & Ito, 2017). Alternatively, smart policy initatives may make sure AI serves a fair, effective, and rewarding function in schools. It is clear, as Selwyn (2019) points out, that AI will impact education in profound ways, so the main concern is how we will help guide these changes.

This chapter focuses on this important subject by studying the developing role of AI in higher education from a policy viewpoint, emphasising future strategies. Several dimensions are covered during the analysis, for example, ethics related to algorithms, transparency and accountability; social aspects including gaps in access and internet boundaries; stakeholders' participation in policy design; international alignment of rules; efforts to raise awareness about AI; mechanisms for monitoring; and well-planned investment in infrastructure and both proactive and early regulatory action. As it examines this subject, the chapter repeats that AI should be governed in education by both encouraging tech growth and keeping safety at the forefront; favouring success in schools and maintaining quality education; staying focused on both human and machine roles; and managing issues that span continents, respecting each environment's distinctions. Instead of decisions between these guidelines, we ought to find ways to integrate them into strong policy plans and management systems (Prinsloo, 2020). The actions higher education institutions take will have a major impact on their future growth as well as on how society relates to advanced technology. Here, we lay out a plan for this significant effort, presenting ideas and useful advice for creating governance systems that promote responsible AI for better education, equality and human health.

It also becomes clear from the literature that important research gaps exist. Few studies have tried to understand the long-term effectiveness of government policies. It is also important to improve research on issues that affect policies' implementation process, mainly technical matters, the ability of institutions, and overall cost. So far, there aren't many studies that compare education across a wide range of cultures, economies, and educational systems, since existing studies often focus on institutions and systems from the Global North (Popenici & Kerr, 2017). Even with these gaps, the new research helps build good theories and gain knowledge that feeds into the creation of new government decisions. This chapter expands on the earlier ideas and shares novel aspects on adaptive frameworks for handling AI challenges in the higher education sector.

2 Evolving Ethical, Social, and Technological Challenges

The integration of AI in higher education unfolds against a backdrop of rapidly evolving ethical, social, and technological challenges that demand dynamic policy responses. This section examines these interrelated dimensions, highlighting key concerns that must be addressed through forward-looking governance frameworks.

2.1 Ethical Considerations in Educational AI

The range of ethical challenges in AI used in education is always growing as technology and the ways it is deployed change. This is shown in

Table 1, which marks the main ethical issues that have developed as AI has been adopted in higher education institutions.

Governance needs to change and adapt to meet present and future ethical challenges instead of fixing only existing ones. As Prinsloo (2020) explained on page 143, AI in education needs to be guided ethically in advance, forming systems that will evolve as technology advances. Algorithmic bias shows just how much ethical matters can change. Fundamentally, early policies attempted to increase diversity in the data, and as models progressed, attention moved to how bias appeared in feature selection, architecture, and their usage in different applications (Ekowo & Palmer, 2017). A study carried out by Kizilcec and Lee (2020) discovered that algorithms used for student outcomes prediction in higher education were biased differently in each institution, meaning that governance needs to be adapted to the context. Now, simply disclosing what algorithms are used is not enough; regulators want algorithms to be much easier to interpret and explain. Since such models are being used in admissions,

Table 1 Evolution of Ethical Challenges in Educational AI

Development Stage	Primary Ethical Concerns	Policy Implications
Early algorithmic systems	Data privacy: Basic fairness	Simple consent frameworks, equal access policies
Machine learning integration	Algorithmic bias transparency explain ability	Bias auditing requirements disclosure standards: Explainable AI guidelines
Advanced adaptive systems	Student autonomy, teacher displacement, educational surveillance	Agency preservation frameworks, role transition policies, surveillance limitations
Emerging general AI	Cognitive dependency, value alignment, educational purpose redefinition	Critical thinking safeguards values-based design requirements, purpose preservation policies

evaluation, and the division of resources, policies should set standards that balance what is possible with what is necessary (Kaplan & Haenlein, 2019).

2.2 Social Implications and Digital Inequality

Policy challenges associated with AI in higher education are rapid and unpredictable since society is evolving at a fast pace. Differences in access, use, and outcomes from digital technologies, known as digital inequalities, are important problems that show up in many contexts (Robinson et al., 2020).
Figure 1 shows that these inequalities have different layers in the context of educational AI.

Reich's (2020) study illustrates that differences in equipment at the start often grow into additional gaps in educational outcomes due to how educational technology is used in various ways. Selwyn et al. (2022) note that the gap in how AI is used by different institutions can further support the status quo in higher education. Social challenges call for strategies that address inequalities in various ways at once. Ensuring that everyone has access to information, despite a lack of use or understanding, is not enough when complementary measures for skills and AI are not added (Holmes & Porayska-Pomsta, 2022). We should also remember that effective governing involves understanding how race, socioeconomic background, and place all affect the ways educational AI is used and the outcomes it achieves.

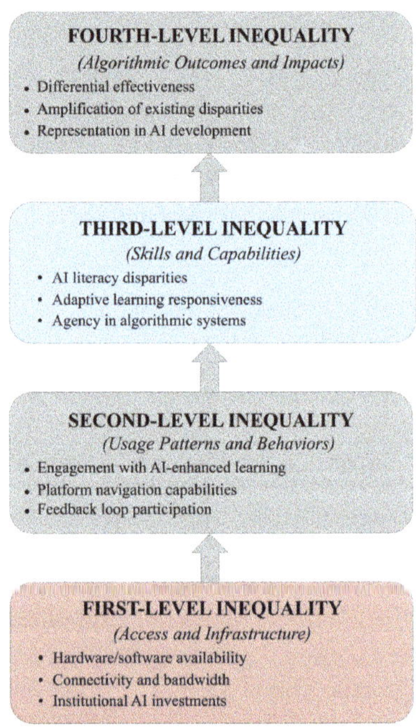

Fig. 1 Multi-level Framework of Digital Inequality in Educational AI

2.3 Technological Evolution and Adaptive Governance

The quick rise in AI abilities is the main difficulty for policymaking today. With technology in this field evolving in unexpected ways with frequent major developments, it becomes very difficult for existing rules to work effectively (Kaplan & Haenlein, 2019). What once were limited educational AI algorithms are now replaced by advanced tools using deep learning, processing speech and language, and predicting results. These new advances offer automatic changes to how instruction is given, materials are made, and student work is evaluated (Holmes et al., 2022). Whenever technology develops, experts have to find new ways to govern and make related policies. This technological dynamism necessitates what Danaher (2019) terms "anticipatory governance" policy frameworks designed not merely to address current capabilities but to adapt proactively to emerging technological developments. Key elements of such adaptive governance include:

(i) Experimental platforms that permit testing new applications.
(ii) A system where reviewing a proposal depends on its level of risk.
(iii) Important milestones should be set for evaluating governance setups based on new technology.
(iv) Technical guidelines that focus on working well together, clear practices, and responsible innovation.
(v) Ethical rules that do not depend on different technologies.

Several institutions are now using some of these adaptive governance ideas. Stanford's framework, created in 2021, includes auto-generated cheques triggered by certain technological breakthroughs, whereas the EUA's initiative, from 2022, builds a collaborative network for adjusting policy.

Because ethics, social impacts, and technological advances meet at this point, policymakers must address these issues with smart solutions. Below, I explore important elements of these replies, first by discussing how to effectively engage stakeholders in developing policy.

3 Stakeholder Engagement and co-Creation of Policies

People who explore, use, and understand AI in higher education should be actively involved in creating AI guidelines. We examine how the process of establishing inclusive policies could improve the acceptance, power, and adaptability of AI governance frameworks.

3.1 Multi-Stakeholder Governance Models

Because of how complicated educational AI is, its policymakers should organize input from multiple communities. It appears that policies developed with the participation of different stakeholders are more complete and succeed better than those decided by only one group (Chan, 2023; Ulnicane et al., 2021). It is shown that building policies by collaborating with various groups works better and that frequent updates and adaptations are necessary to cope with new technologies (Ghimire & Edwards, 2024). You can see the main engagement models compared in Table 2 with regard to higher education AI governance initiatives.

It is revealed in studies (e.g., Kujala et al., 2022) that involving different stakeholders in the planning of policies is more resource-consuming; however, it is more likely to result in sustainable and effective outcomes because all stakeholder concerns

Table 2 Comparative Analysis of Stakeholder Engagement Models in AI Governance

Governance Model	Key Characteristics	Strengths	Limitations	Implementation Examples
Expert-driven	•technical specialists Dominate •limited consultation • efficiency-focused	•technical sophistication •implementation clarity •rapid development	• legitimacy deficits • limited perspective • value blind spots	MIT AI ethics framework Stanford HAI guidelines
Administrative Centered	•institutional leadership control • compliance orientation • risk management focus	•alignment with institutional priorities • resource commitment • implementation authority	• innovation constraints • limited stakeholder buy-in • potential disconnection from Practice	UC Berkeley AI governance Oxford university AI policy
Multi-stakeholder deliberative	• diverse stakeholder inclusion • deliberative processes • iterative development	•enhanced legitimacy •comprehensive perspective •implementation readiness	• time-intensive • resource requirements • potential for compromise dilution	Edinburgh university AI framework University of Helsinki AI Governance
Networked governance	•distributed authority •peer-based accountability •collaborative implementation	• adaptability • knowledge sharing • context sensitivity	• coordination challenges • variable implementation • accountability gaps	European university association AI network Global learning consortium AI principles

are addressed from the start instead of being dealt with later. Involving groups like civil society, government, academia and industry in multi-stakeholder processes can lead to both temporary solutions and those that last in the future, as discussion platforms are provided where stakeholders with a range of opinions can discuss problems and try to solve them, merging scientific findings with another knowledge (Norström et al., 2020). Research shows that policies formed through involving many stakeholders are more widely supported and achieve better results than policies formed through usual administrative practice.

3.2 Inclusive Decision-Making Processes

How different groups participate in policy development shapes the results, more than just the formal rules alone. Real engagement happens when procedural rules ensure that various groups, especially those normally left out of technology decisions, have their voices heard. (Sánchez-Monedero et al., 2020).

The following things should be an important part of any inclusive decision-making process:

(i) Specific approaches to engagement that consider the different skills, time, and preferred methods of communication within every group
(ii) Processes that encourage people to share ideas and arrest the simple counting of likes and disagreements.
(iii) Making clear how stakeholder views lead to certain policy actions.
(iv) Such projects are created to give various participants the important knowledge they need to be part of the movement.
(v) Systems that keep all stakeholders involved both during the process and in the evaluation process.

Research demonstrates that comprehensive stakeholder engagement processes involving diverse participation methods lead to more robust and implementable governance frameworks (Stilgoe et al., 2013). Studies of multi-stakeholder AI governance initiatives show that frameworks developed through inclusive processes that combine technical expertise with broader community input tend to achieve better balance between innovation enablement and risk mitigation (Jobin et al., 2019).

3.3 Addressing Digital Inequality through Participatory Governance

Participatory governance is a key factor in treating digital inequality in the field of educational AI. (Noble, 2018; Eubanks, 2018) found that not engaging marginalised communities in policy development leads to insufficient improvements in promoting fairness in AI. Alternatively, when inclusion is present in governance, it becomes possible to handle and anticipate possible unequal issues before anything is put into practice (Parthasarathy & Katzman, 2024).

Several effective Methods have appeared to make equity a centerpiece in AI governance.

(i) Assessments that thoroughly check whether different student groups are affected in the same way.
(ii) Committees made up of members with different demographics and life experiences.
(iii) Including communities in the research activities that assess different types of technology.
(iv) Ensuring that AI rewards students and communities who have been underrepresented is possible by creating distribution systems which The City University of New York's "AI Equity Initiative" does by building a system that includes underrepresented community members and students in overseeing AI applications (Nelson, 2022).

Observations from the first phase suggest that this process has clarified and corrected equity matters that other governance procedures overlooked. Developing guidelines for AI at universities should focus on active participation by all stakeholders, rather than seeing them just as people to consult (Crawford, 2021). Policy outcomes are effective and accepted only if a range of perspectives is heard from those who take part in formulating governance methods.

4 International Collaborations and Global Standards

Since AI technologies are used, created, and impact different countries, it is necessary to find governance methods that work across both societies and states. Here, we analyse how countries join together to develop effective AI rules for higher education, covering both the benefits and the obstacles.

There are special international difficulties in AI policy development for education that need to be tackled. These include:

(i) Differences in rules between regions when it comes to data protection, accountability around algorithms, and standards for education.
(ii) Crossing borders with data often means that educational information is under he jurisprudence of various and at times conflicting rules.
(iii) International markets are now mainly driven by technological companies working in different countries.
(iv) Many opinions about suitable AI applications in education, from culture and ethics.
(v) Differences in resources and abilities for governance among institutes in the Global North and the Global South.

In internationalised environments, these problems make it hard for universities to coordinate their activities. According to a study of 87 universities from 23 countries, 76% found significant challenges in global AI use, related to keeping data under their control, complying with the rules of different countries, and ensuring standards of ethics are alike (International Association of Universities, 2022).

4.1 Models for International Policy Coordination

Facing these problems, several innovative methods have been developed to help with international coordination. Table 3 explores important methods for international teams to work together on AI in education.

Findings suggest that by using elements from different coordination models, we can better manage the difficult problems involved in managing AI in education (Aiken, 2022). The European University Association has developed the Responsible AI in Higher Education initiative, which combines soft law, network rules, and

Table 3 International Coordination Models for Educational AI Governance

Coordination Model	Primary Mechanisms	Strengths	Limitations	Notable Examples
Harmonized regulation	Legal frameworks with transnational applicability	• regulatory certainty • compliance efficiency • strong enforceability	• sovereignty tensions • implementation complexity • adaptation challenges	EU AI act (educational provisions)GDPR impacts on educational data
Soft law instruments	Non-binding guidelines, principles, and standards	• flexibility • multi-stakeholder development • contextual adaptability	• enforcement challenges • variable implementation • coordination difficulties	UNESCO recommendation on AI ethics OECD AI principles for education
Network governance	Collaborative associations among institutions and stakeholders	• knowledge sharing • best practice development • implementation support	• resource intensity • coordination costs • potential fragmentation	Global higher education AI consortium international AI campus network
Technical standards	Interoperability specifications and development protocols	• practical implementation guidance • technical consistency • market alignment	• narrow technical focus • private sector dominance • limited ethical integration	IEEE learning technology standards IMS global AI interoperability framework

technical standards to form a valid structure for coordinating AI in higher education across countries (European University Association, 2022).

4.2 Balancing Global Standards with Contextual Adaptation

There is a major issue in international governance that involves how to weigh the gains from having global standards against the requirement to adjust for local needs. Even though using the same rules and guidelines allows various countries to cooperate and avoid unfair competition, the quality of governance must still address distinct educational systems, important cultural customs, and varying educational resources (Dale, 2005; Lingard & Rizvi, 2010; Marginson, 2006). Figure 2 highlights a method of running international governance that keeps key beliefs and allows for local responses.

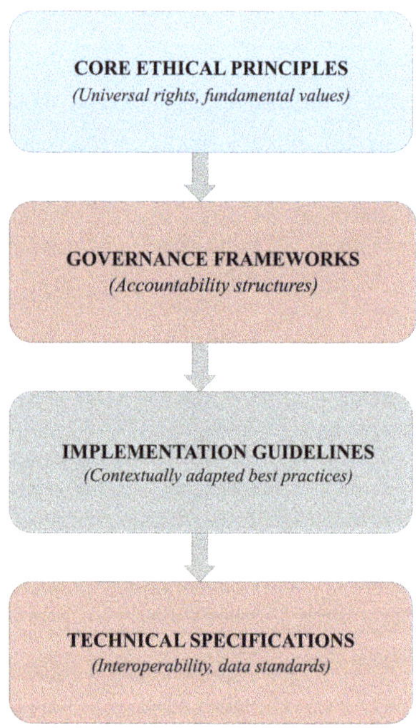

Fig. 2 Layered Model for International AI Governance in Higher Education

4.3 *Case Study: International Observatory on AI in Higher Education*

Effective cooperation among nations can be seen in the launch of the "International Observatory on Artificial Intelligence in Higher Education" in 2022, thanks to a partnership between UNESCO and IARU.

Association of Universities and regional university associations operating on six continents. Several mechanisms in the Observatory foster efficient governance coordination.

(vi) Systems for tracking and identifying distinctions in government approaches worldwide.
(vii) Successful implementation methods are shared and documented in best practice repositories.
(viii) Shared development of governance that strengthens in limited-resource areas.
(ix) Monitoring programmes that report similar results about the use of AI
(x) These dialogues gather various participants into discussions about governance.

Initial findings suggest that the Observatory has greatly improved how participating institutions coordinate policies, especially for issues concerning data governance, the transparency of algorithms, and setting ethical standards (UNESCO, 2023). With AI spreading to many nations as it advances, it is more important than ever for countries to coordinate their actions to steer AI in the best direction. While working with partners, education institutions need to preserve their ability to adapt to their local contexts, student audiences, and what they hope to achieve in education.

5 Integrating AI Literacy and Ethics Education

It is not only about making formal guidelines but also about allowing stakeholders to understand modern technology better. In this section, we study ways to teach AI literacy and ethics to students and professors, as proper understanding is important for everyone in governing AI.

5.1 *Conceptualizing AI Literacy for Higher Education*

Having AI literacy involves knowing about, practicing with, and critically thinking about intelligent technologies. While you work on making code and using algorithms, you should also develop ethical skills, analyse systems, and become good at asking critical questions (Ng et al., 2021). In Table 4, AI literacy is discussed in higher

Table 4 Multidimensional AI Literacy Framework for Higher Education

Literacy Domain	Key Competencies	Educational Approaches	Assessment Indicators
Technical understanding	• algorithmic fundamentals • data literacy • system capabilities and limitations • Interface navigation	• technical modules • hands-on projects • guided experimentation • code exploration	• capability demonstration • application development • system interaction • technical explanation
Ethical reasoning	• value identification • ethical framework application • dilemma analysis • responsible design principles	• case studies • ethical simulations • framework application • value tension exploration	• dilemma resolution • framework application • value articulation • design evaluation
Critical assessment	• bias identification • limitation recognition • impact analysis • alternative consideration	• critical analysis exercises • system interrogation • comparative evaluation • impact assessment	• critique articulation • limitation identification • alternative proposal • impact prediction
Sociotechnical analysis	• context consideration • stakeholder mapping • power relationship analysis • institutional impact	• contextual analysis • stakeholder interviews • institutional ethnography • impact scenarios	• context articulation • stakeholder mapping • prediction power analysis • impact
Practical application	• appropriate tool selection • effective system utilization • output interpretation • adaptation strategies	• applied projects • problem-based learning • domain application • adaptation exercises	• tool selection rationale • effective utilization • result interpretation • adaptation demonstration

education using a framework that highlights various abilities in technology, ethics, and society as the main components.

It involves looking at more than the technical aspects of AI to improve a person's ability to engage with smart technologies in different situations. Ng et al. (2021) discovered that multidimensional frameworks give learners a better view of technology than approaches that look only at skills or ethics.

5.2 Curricular Integration Strategies

AI literacy should be taught in college by strategies that unite detailed, specific knowledge with broad skills in technology use. Several promising methods have appeared lately.

(i) All students should have basic AI knowledge, and core programming courses help them obtain this.

(ii) Topics that go over how AI is applied to different fields.
(iii) Projects involving several fields to look at the various effects of AI.
(iv) Additional workshops, seminars, and project opportunities are offered outside the normal academic courses.
(v) Having students work with AI systems themselves by using them in practice.

Elements of AI combine a general module for all undergrads together with practical AI lessons in various departments and hands-on tests in the AI laboratory (Ng et al., 2021). Program evaluations have demonstrated impressive growth in student AI literacy in technical, ethical, and critical evaluation areas.

5.3 Professional Development for Educators and Staff

AI literacy initiatives at institutions are strongly supported by the development of faculty and staff. In addition to knowing how to use AI, educators need help thinking about the impacts of AI on their work, curriculum, and students (Holmes et al., 2022). Most successful professional development programmes have all of the following:

(i) Essential information on the strengths and weaknesses of AI in practice.
(ii) Uses of AI in teaching practices and relevant case studies.
(iii) Theoretical systems analysing the value effects of using AI in education.
(iv) Teachers should be taught how to get involved when shaping educational policies.
(v) Teachers should evolve their skills as they work in conjunction with smart technology.

Imperial College London has established a structured "AI in Education Faculty Development Programme," which allows staff members to move from gaining the basics in AI to advanced expertise in applying it in education (Imperial College London, 2022). The results of programme evaluations suggest that faculty members grow their technological skills while also gaining a stronger understanding of using technology appropriately.

5.4 Building Cultures of Responsible Innovation

Besides providing formal education, these organisations should encourage wider cultures that respect responsible progress in technology. Such cultural development means:

(i) A set of clearly written values detailing the principles AI will be adopted.
(ii) Two Groups of educators engaged in continuous reflection about new technologies.
(iii) Systems that recognise individuals who help promote responsible innovation.

(iv) Ways for people to share ideas about how technology is changing the world.
(v) Steps are being taken to gather insights from what is implemented and then teach others about them.

At Carnegie Mellon University, the "Responsible Computing Community" project sets up networks across the institution for people to talk, think together, and participate in sharing governance (Carnegie Mellon University, 2022). The project has helped build a way of thinking about technology in the agency that guides the development of rules and their application by staff.

6 Auditing and Monitoring AI Systems

Frameworks for AI should provide regular ways to assess the technology so that it stays in line with what is correct and required by both society and law. This part analyzes techniques for auditing and monitoring the use of AI in education, relating to the use of certain methods, measures and standards of accountability.

6.1 Comprehensive Audit Frameworks

AI audit looks at many aspects, including how well intelligent systems work, how ethical they are, how they can teach, and how they follow rules and standards. The framework in Table 5 connects each of the four areas in a unified way.

Regular reviews taking account of the full guidelines enable organisations to detect possible issues right away. Proactive auditing, as researched by Raji et al. (2020), greatly cuts the chances of algorithmic harms and at the same time boosts performance and trust in the system Table 6.

6.2 Metrics for Evaluating Educational AI Effectiveness

Finding the right ways to measure results is a major challenge within educational AI governance. For an evaluation to be effective, it must measure important educational results, widespread usefulness and ethical practices as well as technical aspects. A balanced scorecard method for evaluating educational AI is demonstrated in Fig. 3.

It makes sure evaluation looks at a range of system factors rather than only favoring technical aspects above learning and values. Each grouping of metrics within a quadrant should be adjusted to suit a particular case, but the method of scoring must stay the same. Longitudinal assessment is vital in education AI since the most important developments happen over a long time. Established in 2023, the Carnegie Mellon University "Educational AI Longitudinal Observatory" shows how such impacts

Table 5 Comprehensive Educational AI Audit Framework

Audit Dimension	Key Considerations	Evaluation Methodologies	Documentation Requirements
Technical performance	• accuracy • reliability • security • robustness	• benchmark testing • adversarial evaluation • performance monitoring • security assessment	• accuracy metrics • reliability reports • security analysis protocols • robustness
Fairness and equity	• demographic performance parity • representation in development • accessibility features • bias mitigation	• disaggregated performance analysis • representation assessment • accessibility evaluation • bias testing	• demographic performance data • development team composition • accessibility compliance • bias mitigation strategies
Transparency and explain ability	• model documentation • decision explanation capabilities • transparency for stakeholders • technical interpretability	• explanation quality assessment • stakeholder comprehension testing • documentation review • interpretability analysis	• model documentation • explanation examples • stakeholder materials • interpretability mechanisms
Educational impact	• learning outcome effects • pedagogical alignment • student engagement • faculty experience	• outcome comparison studies • pedagogical assessment • engagement analysis • faculty interviews	• learning outcome data • pedagogical alignment documentation • engagement metrics • faculty experience reports
Privacy and data governance	• data minimization • consent mechanisms • processes data security • governance	• data flow analysis • consent procedure review • security testing • governance evaluation	• data inventory • consent documentation • security structures certifications • governance
Regulatory compliance	• legal requirements • institutional policies • guidelines industry standards • ethical	• compliance review • policy adherence assessment • standards verification • ethics evaluation	• compliance documentation • policy alignment evidence • standards certification • ethics review results

can be identified by tracking several cohorts over a longer period than short-term evaluations usually show (Carnegie Mellon University, 2023).

Table 6 Strategic Infrastructure Requirements for Educational AI

Infrastructure Dimension	Key Components	Implementation Considerations	Investment Priorities
Technical infrastructure	• computing resources • data systems • network capabilities • security architecture	• scalability requirements • integration challenges • sustainability concerns • security imperatives	• cloud computing frameworks • data governance systems • secure API environments • privacy-preserving architectures
Human capabilities	• technical expertise • educational design skills • governance competencies • implementation capacity	• talent development pathways • role transition challenges • collaborative capabilities • knowledge management	• professional development programs • strategic hiring initiatives • collaborative team structures • communities of practice
Institutional structures	• governance frameworks • decision processes • accountability mechanisms • collaborative structures	• organizational alignment • process integration • authority structures • cross-functional coordination	• governance committees • decision frameworks • accountability dashboards • collaborative ecosystems
Knowledge resources	• research capabilities • evaluation frameworks • learning repositories • implementation guides	• knowledge generation • information dissemination • experience capture • adaptation support	• research initiatives • evaluation frameworks • knowledge repositories • implementation resources

6.3 Accountability Mechanisms and Governance Structures

Monitoring works best when results are clear and backed up by rules that make those responsible answer for their findings. Essential parts of accountability frameworks are:

Deciding which people will check and report on the evaluation results. Procedures can be established to handle the issues that the company sees. Reviews of evaluation outcomes by several expert groups. It is necessary for systems to demonstrate their operations so that performance is revealed clearly. These structures reveal how evaluation findings impact the choices made in implementation. It proves this principle most clearly by presenting the "AI Accountability Framework," which includes a governance process made of special teams, routine reporting, and set measures for resolving problems (Raji et al., 2020). The auditing process means the audit discoveries are converted into changes the governing team can act on and are not just kept as detailed points.

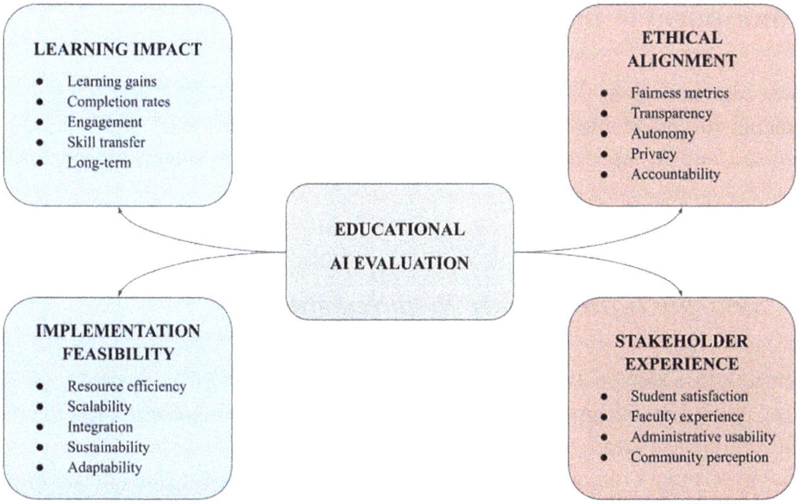

Fig. 3 Balanced Scorecard for Educational AI Evaluation

6.4 *Continuous Improvement Models*

Monitoring, compliance, and auditing of AI should help boost its system's performance and how it is looked after. That's why we need frameworks that:

(i) Make use of the evaluation outcomes to improve the system through software development.
(ii) Let stakeholders' practical knowledge guide the way improvements are built into the design and process.
(iii) Experience in implementing documents helps the organisation gather important information for its knowledge base.
(iv) Viewing practices in actual use allows you to improve how you use them.
(v) Helping each other learn and apply knowledge in different areas can be very beneficial.

By teaming computer experts with educators in a repeat cycle of design and review, ensures that findings are turned into helpful improvements (Luckin & Holmes, 2016). Thanks to this model, AI has achieved a lot more in both technical and educational areas at the institution compared to in the past.

7 Investment in Infrastructure and Research

Higher education considers building and studying infrastructure key elements for advancing future-oriented AI governance. We discuss here what areas should be prioritised, how to shape research, and the best ways to train someone for responsible AI and academic advancement.

7.1 Strategic Infrastructure Requirements

To make educational AI successful, we must invest heavily in systems, people, and organisational design. Table 6 outlines which elements are considered the most significant infrastructure needs in all of the above fields.

The government ought to look ahead while also making sure, here and now, that governance does not become unsustainable. UC follows this equity approach by dividing its funding between various technological and community needs and by always considering what is crucial now and in the future.

7.2 Research Priorities for Equitable

The results of research are the next steps in the growth and use of AI in education. Choosing to focus research on equity, how well it works, and ethical use can help. Have a strong impact on the progress of technology. Several domains in research look especially promising. Working for responsible advancements in technology.

(i) Approaches in building AI systems are developing so that they address rather than worsen educational equity.
(ii) Projects investigating how various schools and teachers implement AI systems.
(iii) Using technology within teaching practices is the focus of many pedagogical integration frameworks.
(iv) Meaningful participation of stakeholders is encouraged by developing new ways to evaluate the system.
(v) The area looks at how different policies impact how well things are carried out in practice.

In 2022, the MacArthur Foundation launched the "Equitable AI in Education Initiative," supporting joint studies that pay specific attention to making technology practical for those who do not necessarily have access to it in education (MacArthur Foundation, 2022). The project has delivered frameworks for equitable design, toolkits for wide use in places with scarce resources, and systems that support active community involvement.

7.3 *Aligning AI Initiatives with Broader Societal Goals*

In higher education, AI governance should make explicit links between technology and sustainability, social justice, democracy, and help people flourish. For this to happen, investment strategies must be:

(i) Connect directly with how technological progress helps the community achieve its aims.
(ii) Related organisations and financial institutions should put in place schemes that incentivize green technologies that help achieve the main objectives.
(iii) Examine how implementation is done by using different criteria.
(iv) Work with partners outside education to use AI to help society
(v) Design procedures that prepare for the future effects on society

Figure 4 helps show how to connect educational AI initiatives to the Sustainable Development Goals.

The initiative named "AI for Social Impact" by the Association of Public and Land-grant Universities (2022) uses this approach by deciding on funding and evaluation methods that relate education AI projects to addressing sustainability and social

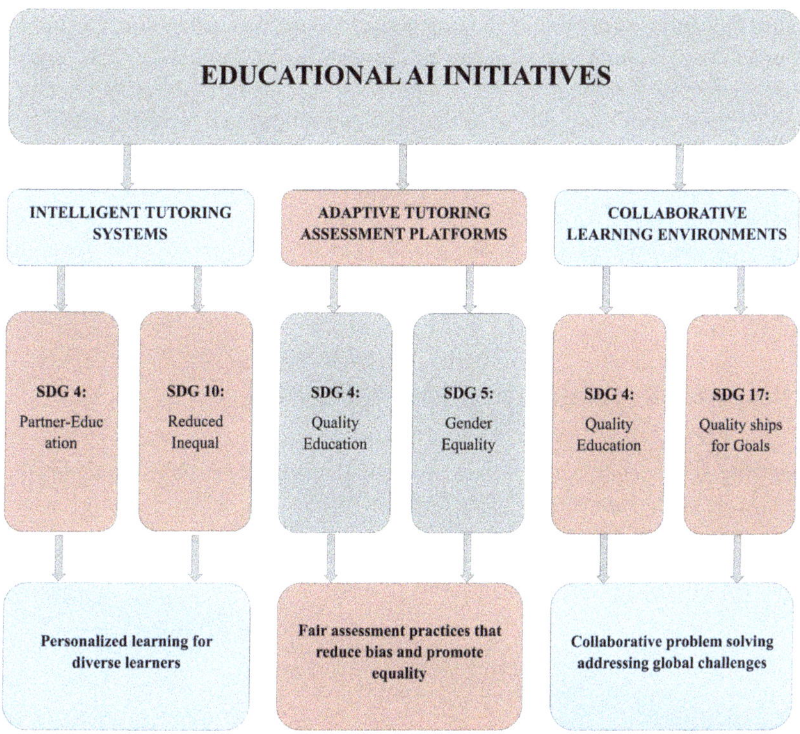

Fig. 4 Aligning Educational AI with Sustainable Development Goals

justice concerns. This programme has promoted the growth of tools that support educational equity, climate education, public health, and opportunities for citizen involvement, also establishing new rules for ethically based innovation.

7.4 Public-Private Partnerships for Responsible Innovation

This type of investment requires public and private entities to team up to benefit from each group's available resources, skills, and insight. Public-private partnerships that work well in managing educational AI usually consist of:

(i) Everyone in leadership agrees on the importance of educational principles and values.
(ii) Defined roles, responsibilities, and who can decide on what.
(iii) Ways to help people learn from each other, ensuring intellectual property rights.
(iv) Structures that oversee the way collaborative programmes operate.
(v) Sustainable ideas that provide for the economy and maintain what education needs.

The partnership approach in AI is proven by the creation in 2023 of the "Education AI Commons" thanks to cooperation among public universities, educational technology companies, and philanthropic foundations (Education AI Commons, 2023). The plan uses public money for research, receives financial support from the private sector for technology, and coordinates the effort so it follows educational priorities and focuses on equality. In this way, they lay down rules for building AI responsibly, which also lead to improvements in both technology and education. This next part looks at the details of each.

The way the development process is shaped depends on suitable governance measures.

8 Proactive Regulation and Risk Mitigation

Overseeing AI in schools calls for regulations that support responsible change and minimize possible side effects. The purpose of this section is to explore ways to include measures in regulations that anticipate problems, lower risks, and adjust to recent advancements in technology.

8.1 Balancing Innovation and Oversight

Building good regulations depends on balancing the need for technological progress with overseeing the industry so that accidents are not common. Setting clear boundaries with regulators can fix this issue and give investors a space to innovate. In Table 7, we summarize the significant features of regulatory frameworks that encourage innovation.

According to Gasser and Almeida (2022), encouraging responsible practices and new technologies goes hand in hand when regulators empower businesses to act responsibly instead of concentrating only on meeting standards. Based on results from different countries, it is clear that flexible rules for regulations perform best in helping manage educational technology matters.

Table 7 Characteristics of Innovation-Enabling Regulatory Frameworks

Regulatory Dimension	Innovation-Inhibiting Approaches	Innovation-Enabling Approaches	Implementation Examples
Regulatory focus	• prescriptive technical requirements • detailed procedural specifications • rigid implementation mandates	• outcome-based standards • principle-centered guidance • performance objectives	• EU AI act educational provisions • UNESCO AI ethics implementation guide • UK AI governance framework
Risk calibration	• uniform requirements regardless of application • binary compliance determinations • static risk assessments	• risk-tiered requirements • proportional oversight mechanisms • dynamic risk evaluation	• Singapore's AI governance framework • Canadian algorithmic impact assessment • OECD risk-based approach
Implementation pathways	• single compliance pathway • limited implementation timeframes • rigid conformity requirements	• multiple compliance pathways • phased implementation • equivalence provisions	• New Zealand AI regulatory sandbox • UK ICO regulatory guidance • Australia's AI ethics framework
Adaptation mechanisms	• fixed regulatory requirements • infrequent revision cycles • limited feedback incorporation	• regular review processes • stakeholder feedback channels • technological evolution provisions	• Finland's adaptive AI governance • Estonian digital governance framework • Singapore's living lab approach
Innovation support	• compliance focus • penalty orientation • guidance: Limited implementation	• capacity building provisions • implementation resources • technical assistance	• EU AI innovation hubs • UK AI standards hub • Canadian AI governance support

8.2 The EU AI Act as a Regulatory Model

The European Union's AI Act, which was finalized in 2023, stands out because it is important for governing educational AI. According to the Act, various AI systems are assigned to specific risk classes and required to comply with matching regulations, a practice now being adopted by other countries (European Commission, 2023). The approach taken by the EU in educational governance covers the main points:

(i) It places AI applications into different groups depending on their possible impact significance.
(ii) These applications tell users what they are not allowed to do on the platform.
(iii) Heavy monitoring is required for requirements that bring about important outcomes.
(iv) Public disclosure of particular AI interactions is required by law.
(v) They are designed to provide a safe space for experimentation with new practices.

When it comes to education, the Act views systems used for making decisions about assessment, admissions and personalised learning to be "high-risk," and sets clear standards for them. They help develop the way education is governed worldwide because they balance introducing new solutions with necessary risk controls.

8.3 Anticipatory Governance and Adaptive Regulation

Because AI technology changes quickly, new rules are needed that can change as well. A regulatory paradigm focused on being ready for upcoming educational AI seems especially useful as it proactively tackles new abilities, rather than only paying attention to current tools (Danaher, 2019).

The framework for anticipatory governance in educational AI contexts is shown in Fig. 5.

8.4 Institutional Risk Mitigation Strategies

Besides government regulations, colleges and universities should create risk management strategies that work for their particular environment. Usually, effective strategies consist of:

(i) AI governance teams consisting of members from various departments and having official responsibility.
(ii) Collected data is studied in protocols before a new implementation can be deployed.
(iii) Procedures outline what is necessary for responsible reporting.

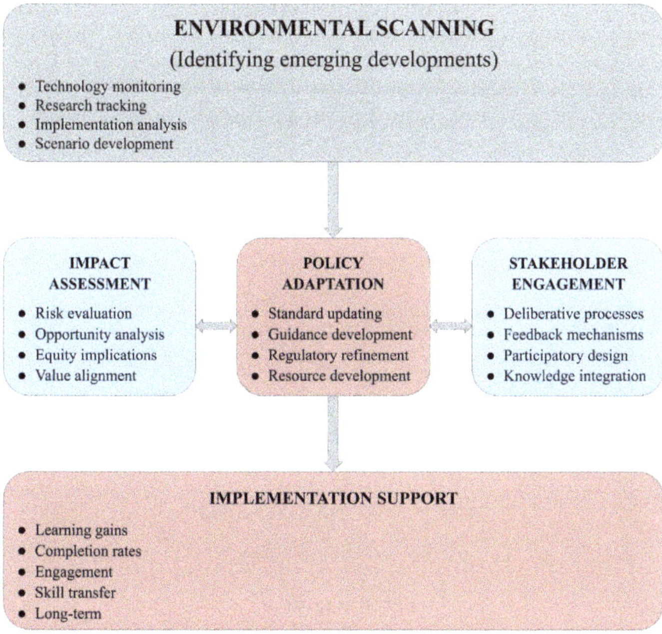

Fig. 5 Anticipatory Governance Framework for Educational AI feedback and flexible pathways to put changes into practice (Masamuddin & Rashid, 2022)

(iv) Frameworks that keep an eye on how the system functions and what outcomes it has.
(v) Incident response procedures are designed to immediately and effectively handle recognised challenges.

This framework covers pathways to manage risk, approve use, regularly check progress and react to incidents at the University of Oxford (University of Oxford, 2025). Learning analytic tools, automatic assessments, and administration decision support are all part of what the framework covers for education. While educational AI grows and changes, managing it well will require a mix of ethical rules and the ability to support new responsible ideas in technology. The most valuable frameworks have properly calibrated standards, reliable assessment systems, and flexible rules that grow with technology, yet always center on educational values.

9 Integrated Policy Framework

Since the findings support this need, Fig. 6 introduces an approach that links the main aspects examined earlier in the chapter.

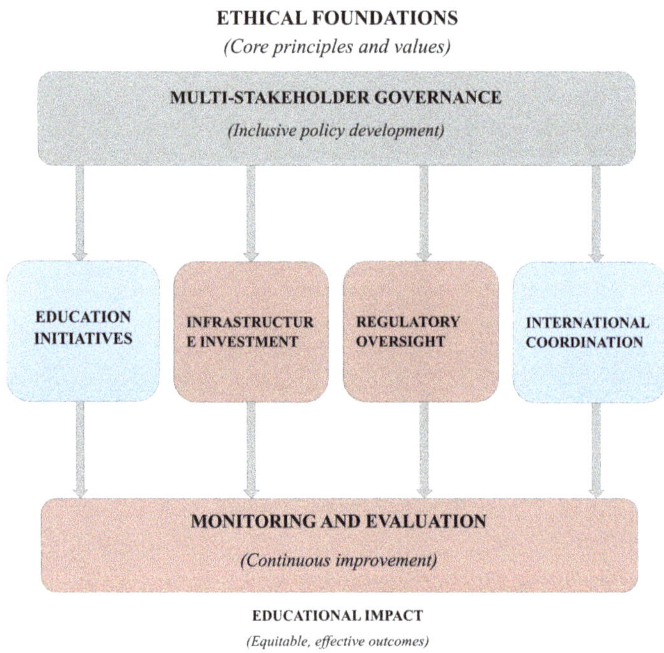

Fig. 6 Integrated Policy Framework for Educational AI Governance

9.1 Recommendations for Future Directions

From the information presented in this chapter, some main recommendations are offered to key players in educational AI management.

(i) Produce policies that adjust to technological progress and remain committed to basics of ethics. They must set out clear limits and help ensure responsibly pioneering technologies with checked and reviewed requirements.
(ii) Adopt procedures for policy development that involve many different stakeholders in an organized way. Making sure communities who have long been excluded from decisions about technology are involved will help to promote equal results.
(iii) Create comprehensive AI programs that teach people about the different parts, ethics and social impacts of intelligent technology. This work should involve both course designs and activities for teachers, efforts to build the institution's culture and engagement with the community.
(iv) Design methods for assessing educational AI in aspects such as student outcomes, ethical behavior, how easily it is implemented and the experiences of individuals connected with it. The main idea of these approaches should be to assist in betterment and not only to verify whether standards are met.

(v) Prioritize investments in infrastructure and research that advance equitable, effective, and ethical implementations. Particular emphasis should focus on applications addressing educational disparities, promoting sustainability, and strengthening democratic participation.
(vi) Participate in worldwide organizations that govern, create standards, and help share knowledge. Such efforts should ensure that one solution works well for everyone while still fitting local needs.
(vii) Use environmental scanning, impact evaluation, and adjustable policies that allow governments to act toward technological changes before they become usual.
(viii) Inspire responsible innovation by setting common values, working in groups that reflect on their actions, creating the right rewards, and using institutional learning to help make policies and informal actions.

10 Conclusion

In conclusion, the integration of artificial intelligence (AI) in higher education holds tremendous potential to transform learning environments, improve administrative efficiency, and enhance student outcomes. However, realizing this potential requires careful consideration of the ethical, social, and technological challenges that accompany AI adoption. As highlighted throughout this chapter, the successful implementation of AI in educational contexts depends not only on technological sophistication but also on thoughtful governance frameworks that prioritize transparency, accountability, and equity. To effectively address these challenges, future AI governance must prioritize adaptable policy frameworks that evolve alongside rapidly changing technological landscapes. This includes proactively addressing issues like data privacy, algorithmic bias, and digital inequality, while ensuring that AI systems align with broader educational values such as academic integrity, inclusivity, and personalized learning. Equally important is the need for ongoing stakeholder engagement, including educators, students, policymakers, and technology developers, to ensure that AI policies reflect diverse perspectives and respond to the needs of all educational communities. International collaboration also plays a critical role in shaping responsible AI policies. By sharing best practices, harmonizing ethical standards, and supporting cross-border knowledge exchange, institutions can create robust governance models that promote equitable access to educational AI technologies. This approach aligns with global frameworks like the EU's AI Act, which emphasizes risk-based regulation and proactive oversight to ensure that AI technologies benefit all stakeholders.

Finally, investment in digital infrastructure, research, and AI literacy is essential for building resilient educational systems. This includes not only technical capacity building but also fostering cultures of responsible innovation through professional development and interdisciplinary research. Regular auditing and monitoring of AI systems, combined with transparent reporting and accountability mechanisms, are also necessary to maintain public trust and ensure long-term success. As higher education institutions continue to navigate the complexities of AI adoption, a balanced approach that integrates innovation with ethical responsibility will be essential. By prioritizing adaptive governance, inclusive decision-making, and sustained collaboration, institutions can unlock the full potential of AI while safeguarding the core mission of education: to empower learners, promote critical thinking, and advance human knowledge.

References

Aiken, R. M. (2022). Cultural dimensions of AI governance in education: A comparative policy analysis. *Journal of Education Policy, 37*(4), 412–431.

Association of Public and Land-grant Universities. (2022). *AI for social impact initiative: Framework and priorities*. APLU Publications.

Bennett Moses, L., & Vellido, A. (2021). Ethical governance of AI in education: A multi-stakeholder analysis. *British Journal of Educational Technology, 52*(6), 2051–2073.

Carnegie Mellon University. (2022). *Responsible computing community initiative: First year report*. CMU Publications.

Carnegie Mellon University. (2023). *Educational AI longitudinal observatory: Research design and early findings*. CMU Press.

Chan, C. K. Y. (2023). A comprehensive AI policy education framework for university teaching and learning. *International Journal of Educational Technology in Higher Education, 20*(1), 38. https://doi.org/10.1186/s41239-023-00408-3

City University of New York. (2022). *AI equity initiative: Framework and implementation guide*. CUNY Press.

Crawford, K. (2021). *The atlas of AI: Power, politics, and the planetary costs of artificial intelligence*. Yale University Press.

CUNY. (2024). Tech equity initiative. *The City University of New York*. https://www.cuny.edu/about/administration/offices/ocip/partners/tech-equity-initiative/

Dale, R. (2005). Globalisation, knowledge economy and comparative education. *Comparative Education, 41*(2), 117–149. https://doi.org/10.1080/03050060500037344

Danaher, J. (2019). Welcoming robots into the moral circle: A defence of ethical behaviourism. *Science and Engineering Ethics, 26*(4), 2023–2049. https://doi.org/10.1007/s11948-019-00119-x

Commons, E. A. I. (2023). *Strategic plan 2023–2027: Collaborative governance for educational AI*. Education AI Commons Publications.

Ekowo, M., & Palmer, I. (2017). *Predictive analytics in higher education: Five guiding practices for ethical use*. New America Foundation.

Eubanks, V. (2018). *Automating inequality: How high-tech tools profile, police, and punish the poor*. St. Martin's Press

European Commission. (2023). Artificial Intelligence Act. *Official Journal of the European Union*.

European University Association. (2022). *Responsible AI in higher education: A framework for European universities*. EUA Publications.

Gasser, U., & Almeida, V. A. (2022). A layered model for AI governance. *IEEE Internet Computing, 21*(6), 58–62.

Ghimire, A., & Edwards, J. (2024). From Guidelines to Governance: *A Study of AI Policies in Education*. ArXiv.org. https://arxiv.org/abs/2403.15601

Holmes, W., & Porayska-Pomsta, K. (2022). The ethics of AI in education: Practices, challenges, and debates. *British Journal of Educational Technology, 53*(5), 1572–1587.
Holmes, W., Saritas, O., Bialik, M., & Bakhshi, H. (2022). Anticipatory futures for AI in education. *Learning, Media and Technology, 47*(4), 495–515.
Imperial College London. (2022). *AI in education faculty development program: Framework and outcomes*. Imperial College Publications.
International Association of Universities. (2022). *Global survey on AI governance in higher education: Findings and implications*. IAU Publications.
International Association of Universities. (2023). *International observatory on AI in higher education: First annual report*. IAU Publications.
Jobin, A., Ienca, M., & Vayena, E. (2019). The global landscape of AI ethics guidelines. *Nature Machine Intelligence, 1*(9), 389–399. https://doi.org/10.1038/s42256-019-0088-2
Kaplan, A., & Haenlein, M. (2019). *Siri, Siri, in My Hand Who's the Fairest in the Land On the Interpretations, Illustrations, and Implications of Artificial Intelligence. Business Horizons, 62*, 15–25. References Scientific Research Publishing. (2019). Scirp.org. https://www.scirp.org/reference/referencespapers?referenceid=2918544
Kizilcec, René F, & Lee, H. (2020). *Algorithmic Fairness in Education*. ArXiv.org. https://arxiv.org/abs/2007.05443
Kujala, J., Sachs, S., Leinonen, H., Heikkinen, A., & Laude, D. (2022). Stakeholder engagement: Past, present, and future. *Business & Society, 61*(5), 1136–1196. https://doi.org/10.1177/00076503211066595
Lingard, B., & Rizvi, F. (2010). *Globalizing education policy*. Routledge.
Luckin, R., & Holmes, W. (2016). Intelligence unleashed: An argument for AI in education.
MacArthur Foundation. (2022). *Equitable AI in education initiative: Program framework and funding priorities*. MacArthur Foundation Publications.
Marginson, S. (2006). Dynamics of national and global competition in higher education. *Higher Education, 52*, 1–39. https://doi.org/10.1007/s10734-004-7649-x
Masamuddin, A. A. B., & Rashid, M. F. (2022). Function and hierarchy of settlement in regional development corridor Malaysia: A case study of South-East Pahang Development Authority (DARA). *The Indonesian Journal of Planning and Development, 7*(2), 84–89.
Nelson, A. R. (2022). Higher education and human capital and in the "New York Bay Area": Historical lessons from the City University of New York (CUNY). In *Higher education, innovation and entrepreneurship from comparative perspectives: Reengineering China through the Greater Bay economy and development* (pp. 17–58). Springer Nature Singapore.
Ng, D. T. K., Leung, J. K. L., Chu, S. K. W., & Qiao, M. S. (2021). Conceptualizing AI literacy: An exploratory review. *Computers and Education: Artificial Intelligence, 2*, 100041. https://doi.org/10.1016/j.caeai.2021.100041
Noble, S. U. (2018). *Algorithms of oppression: How search engines reinforce racism*. NYU Press.
Norström, A. V., Cvitanovic, C., Löf, M. F., West, S., Wyborn, C., Balvanera, P.,... & Le Tissier, M. (2020). Principles for knowledge co-production in sustainability research. *Nature Sustainability, 3*(3), 182–190. https://doi.org/10.1038/s41893-019-0448-2
Parthasarathy, S., & Katzman, R. (2024, December 17). Bringing communities in, achieving AI for all. *Issues in Science and Technology*. https://issues.org/artificial-intelligence-social-equity-parthasarathy-katzman/
Popenici, S. A. D., & Kerr, S. (2017). Exploring the impact of artificial intelligence on teaching and learning in higher education. *Research and Practice in Technology Enhanced Learning, 12*(1). https://doi.org/10.1186/s41039-017-0062-8
Prinsloo, P. (2020). Of black boxes and algorithmic decision-making in (higher) education–a commentary. *Big Data & Society, 7*(1), 205395172093399. https://doi.org/10.1177/2053951720933994
Raji, I. D., Smart, A., White, R. N., Mitchell, M., Gebru, T., Hutchinson, B., Smith-Loud, J., Theron, D., & Barnes, P. (2020). *Closing the AI Accountability Gap: Defining an End-to-End Framework for Internal Algorithmic Auditing*. ArXiv.org. https://arxiv.org/abs/2001.00973

Reich, J. (2020). *Failure to Disrupt*. https://doi.org/10.2307/j.ctv322v4cp
Reich, J., & Ito, M. (2017). *From good intentions to real outcomes: Equity by design in learning technologies*. Digital Media and Learning Research Hub.
Sánchez-Monedero, J., Dencik, L., & Edwards, L. (2020, January). What does it mean to 'solve' the problem of discrimination in hiring? Social, technical and legal perspectives from the UK on automated hiring systems. In *Proceedings of the 2020 conference on fairness, accountability, and transparency* (pp. 458–468).
Selwyn, N. (2019). *Should Robots Replace Teachers? Polity*. https://research.monash.edu/en/publications/should-robots-replace-teachers-ai-and-the-future-of-education
Selwyn, N. (2022). The future of AI and education: Some cautionary notes. *European Journal of Education, 57*(4), 620–631. https://doi.org/10.1111/ejed.12532
Stilgoe, J., Owen, R., & Macnaghten, P. (2013). Developing a framework for responsible innovation. *Research Policy, 42*(9), 1568–1580. https://doi.org/10.1016/j.respol.2013.05.008
Ulnicane, I., Eke, D. O., Knight, W., Ogoh, G., & Stahl, B. C. (2021). Good governance as a response to discontents? Déjà vu, or lessons for AI from other emerging technologies. *Interdisciplinary Science Reviews, 46*(1–2), 71–93. https://doi.org/10.1080/03080188.2020.1840220
UNESCO. (2023). *Artificial intelligence and education: Guidance for policy-makers*. UNESCO Publishing.
University of Oxford. (2025). *AI governance. Oxford Martin School*. Retrieved June 1, 2025, from https://www.oxfordmartin.ox.ac.uk/ai-governance
Zawacki-Richter, O., Marín, V. I., Bond, M., & Gouverneur, F. (2019). Systematic review of research on artificial intelligence applications in higher education–where are the educators. *International Journal of Educational Technology in Higher Education, 16*(1). https://doi.org/10.1186/s41239-019-0171-0

AI in Higher Education: A Catalyst for Sustainable Development

Renjith V. Ravi

Abstract Democratization and proliferation of Artificial Intelligence (AI) in higher education continues to transform new landscapes, providing pathways to accelerate the attainment of the United Nations Sustainable Development Goals (SDGs), notably, SDG 4: Quality Education. This review argues how AI can be used as a tool to improve educational access, equity, and quality towards achieving a more inclusive and sustainable multi-faceted education. The employment of AI technologies in higher education addresses various challenges like educational inequities and quality improvement, development of learner-centred education, enhancing managing support, and other educational processes. Also, AI that promotes lifelong education and especially assists the different underprivileged groups is altogether relevant to the mission promoting high quality of education globally. But the issue of ethics regarding AI, in particular data protection, algorithmic equality, and the distribution of AI technology, also need to be considered. To promote responsible and equitable practice, it has to be addressed. This review also reviews existing literature and practical case studies of educational innovations including AI innovations which are already contributing to the attainment of SDG 4 and presenting opportunities for utilizing AI in the quest for development goals in higher education institutions.

Keywords Artificial intelligence · Higher education · SDG 4 · Quality education · Sustainable development goals · Personalized learning · Educational equity · Lifelong learning · Ethical AI · Education technology (EdTech)

1 Introduction

The role of Artificial Intelligence (AI) in higher education is transforming education organizations across the globe (Yadav, 2024). Powered technologies on AI can transform education by offering personalized learning experiences, automating repetitive

R. V. Ravi (✉)
School of Applied Sciences Engineering and Technology, Rashtriya Raksha University, Gandhinagar, Gujarat, India
e-mail: renjithravi.research@gmail.com

A. Ara (ed.), *Role of Artificial Intelligence in Higher Education*,
https://doi.org/10.1007/978-981-95-5350-1_7

steps, and broaden the scope of education to get to more diverse young people. As we, (the global community), all try our best to achieve the Sustainable Development Goals (SDG) of United Nations, especially SDG 4: Quality Education, AI is being emerged as an enabler for closing educational gap, improving teaching methods, and enhancing inclusiveness. Here, we will highlight the significance of AI in learning and teaching, its contribution to sustainable development, and what specified targets of SDG 4 are being responded to by AI knowledge-based initiatives (Mandal, 2020).

1.1 Importance of AI in Education

Higher education has become an advanced higher education composed of AI technologies. Adaptive learning using AI-powered applications allows the students to get customized content according to their strength and weakness (Darıcan, 2025). Student progress is analyzed by machine learning algorithms for the purpose of providing targeted interventions by educators. Round the clock support is given to students through virtual assistants and AI powered tutors who ensure that students get immediate guidance. In addition, AI based predictive analytics help educators anticipate at risk students to provide remedial help for high retention and graduation rate.

Apart from the practical applications of AI in direct learning, it also improves the administrative efficiency. AI driven chatbots are used by universities in place of humans to address student queries in their departments which help in bringing down the administrative workload and response time. The use of automated grading systems helps to reduce the time taken by faculty in grading papers while reducing their workload, thus facilitating this group to work on the development of curriculum and students engagement. Plagiarism detection tools based on artificial intelligence enable them to detect the unoriginal content in a more accurate way upholding the academic integrity.

All of these (not to mention concerns about algorithmic biases and ethical issues) are challenges of applying AI to education. On the other hand, while designed and applied properly, AI can democratize education by levelling playing field for learning, bringing quality education to all over the world, and in a larger context, contributing to sustainable development.

1.2 The Role of AI in Sustainable Development

Based on the United Nations' definition of sustainable development, it is the process whereby development doesn't go beyond what people need now but also leaves future generations with the capacity to meet their own needs (Niposium, 2023). For all, inclusive, equitable and quality education is in the 2030 Agenda for Sustainable

Development which is comprised of 17 SDGs, with SDG 4 being the one that highlights the importance of the need to ensure. It is in light of this goal that AI has greatly come of age in addressing the issue of education opportunities, improving learning outcomes, and supporting inclusive and lifelong learning efforts.

AI driven technologies ensure availability of education to people in remote and less serviced regions by helping them have access to the education. Translation and speech recognition based on AI technology remove language barriers that allow kids growing up in different languages to consider quality education (Kaushik et al., 2024). Moreover, with AI it is possible to personalize educational experience and students with disabilities receive specially designed support for education and cultivate inclusivity.

There are also aspects of AI that contribute in improving the education quality by making a data-driven decision. AI analytics help universities and policymakers to check the efficiency of education programs and the trends of learning, as well as to allocate the resource more effectively. Whereas AI driven simulations and virtual reality (VR) solutions allows students to engage in immersive learning experience through complex concepts in a way that progressed and ancient studies could not.

Besides, AI is also helpful for lifelong learning with the provision of the flexible and personal learning channels available for individuals who wish to up or re skill to an evolving job market. Using AI-driven recommendation system, courses suggested by the system are the ones most relevant for the particular learning patterns of the learner to ensure that he remains engaged, learning new things and to acquire new things that can lead him to his career aspirations. This is aligned with the SDG 4's target to promote lifelong learning opportunities for all, and education should not be confined to the traditional academic setting.

1.3 SDG 4: Quality Education and Its Targets

And SDG 4 is to "ensure inclusive and equitable quality education (Dawson, 2021) and promote lifelong learning opportunities for all." Meanwhile, AI driven innovation projects fit this goal by targeting some of the main targets of SDG 4.

4.1—Free, equitable and quality primary and secondary education—Using free and high-quality educative platforms and materials powered by Artificial intelligence, they can offer free education to students from disadvantaged backgrounds locations at no cost.

4.2.—Early childhood development and pre-primary education—AI powered educational games and interactive learning tools helps in promoting Early childhood development as well as providing access to basic literacy skills.

4.3—Equal access to affordable technical, vocational, and higher education: AI enabled MOOCs (Massive Open Online Courses) and other online learning platforms like Coursera and edX offer affordable education program to learners across the globe and help in getting equal access to higher education.

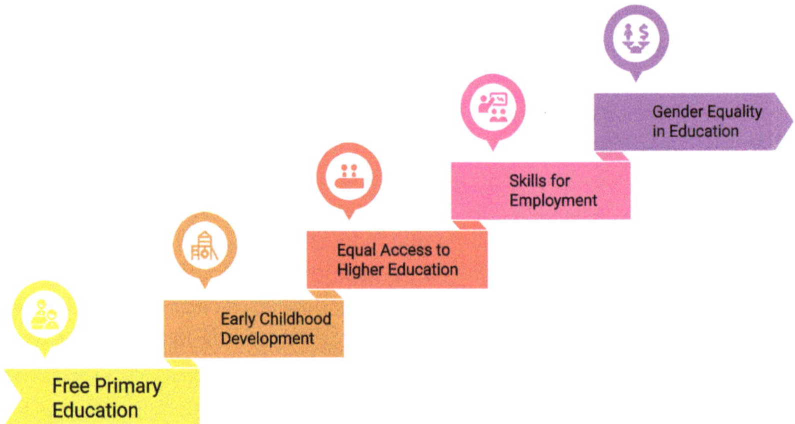

Fig. 1 Pictorial representation of targets in SDG 4

4.4—Increase in relevant skills for employment and entrepreneurship—AI powered career guidance system and individualized and sequenced learning pathways enable students to acquire those skills that are required for the jobs, aligned to what is needed in the real world.
4.5—Elimination of gender disparities in education: Maximum use of current AI-based initiatives like mobile learning apps and AI assisted mentorship progams for the education of women and under representatives, enabling them with the educational resources as well as career development opportunities.
4.6—Universal literacy and numeracy: AI based literacy application like intelligent tutoring system aimed to develop universal literacy and numeracy.
4.7—Education for sustainable development and global citizenship—AI empowered simulations and gamified learning experience help students to realise basic knowledge and skills on the challenges of sustainability and to contribute to sustainable development.

A pictorial representation of all these targets is depicted in Fig. 1

With the help of AI, higher education institutions can play an active role in reaching these goals, and then keep education solidly placed on the pedestal of sustainable development.

1.4 Challenges and Ethical Considerations

There are several opportunities for how AI can improve higher education and increase SDG 4, but some challenges and ethical concerns should be taken into account to implement it properly (Dangol, 2024). Key challenges include:

- It depends on a great deal of student data, often representing a Data Privacy and Security concern. To gain trust of the people in AI driven education systems, it is very important to ensure data privacy and security of student's record and comply with data protection regulations.
- Algorithmic Bias and Fairness: The AI model will learn from the biases in the training data and result in discrimination. Fairness aware AI models must be implemented in institutions to eliminate biases and maintain fair learning experiences to all the students.
- Digital Divide and Accessibility: On the one hand, AI can make accessibility easier, but on the other, it makes it worse in the context of the digital divide: If access to AI utilities is restricted to privileged institutions. Investing in AI infrastructure as well as digital literacy initiatives is essential to bridging the digital divide and making education inclusive.
- Ethical principles must be followed when implementing the AI otherwise it can be dangerous and unethical as well. Creating AI governance frameworks and ethical building guides are two of the methods that can lead to accountable AI deployment in education.

There are many possibilities on how the AI can actually revolutionize the higher education, including making the learning process more accessible, personalized, efficient, etc. Colleges and state rulers can lead to the availability of comprehensive and fair instructive open doors that empower individuals the whole way across the globe by synchronizing AI driven advancements with SDG 4. Nevertheless, this adoption demands, among other things, ethical tensions, privacy of data, and closing digital gaps to be tackled (Villarino, 2024). How AI will play a very crucial role in the future of higher education in regards to a sustainable and inclusive learning future. Investment in AI algorithmic education and AI ethical governance via collaborative effort among higher education institutions, using the full capacity of AI, can make the achievement of SDG 4 possible for said education institutions while also being able to contribute to global sustainable development.

2 AI's Role in Enhancing Educational Access and Equity

It can improve accessibility to higher education and the width of the ladder of learning opportunities through Artificial Intelligence (AI) (Fathima Shah, 2024). SDG 4 is one of the global agenda for sustainable development which stresses on the need to provide inclusive and equitable quality education for all. No section of society is immune either to technology or to its capabilities, especially since AI technologies provide solutions that can solve education disparity, personalize learning and support the underprivileged populations. This section explores how AI can be used to improve educational access by addressing the associated innovations, real world implementations as well as challenges that need to be addressed.

2.1 *AI-Driven Solutions for Educational Access*

It is disrupting the way students now learn using educational resources and shattering the common barriers associated with geography, socio economic background and also physical disabilities (Lowa, 2024). Key AI-driven solutions include:

- **AI Powered Virtual Learning Platforms**: They allow students from the remote areas to avail high quality education without the need for being present in the classroom. These platforms can become adaptive learning hubs thanks to the use of AI; they will offer students learning experiences suited to their pace and level of understanding (education available AND education is effective).
- **Language translation and speech recognition with the help of AI**: These include Google Translate, Microsoft translator and others, which help in breaking barriers of different languages used by students for accessing education material in varied languages. It is particularly useful in international students and multilingual course learning environments where, being able to tackle linguistic content can contribute to a more ultimately personalized learning experience.
- **AI based Accessibility Tools**: Speech-to-text converters, screen readers and personalized assistive technology tools enabled with AI Applications helps the students with disabilities to have the equal access on their education. These tools help in making inclusivity by offering personalized learning helps for understudies who have visual, audio or cognizant handicaps.
- **Intelligent Tutoring Systems (ITS)**: AI driven tutoring system like Carnegie Learning's MATHia and IBM Watson Tutor for example provide customized instruction and real time feedback by adapting to the individual learning pattern and helps the students master complex concept in an effective way.
- **Learning analytics are powered by AI**: AI can analyze huge volumes of recorded performance data of students and recommend personalized learning paths suggesting timely intervention as well as support to students. A tabular summary of these solutions are listed in Table 1.

These are the AI powered innovations that are fundamental in ensuring that students located in areas with limited socio-economic development as well those in remote rural areas have access to education.

2.2 *Addressing Educational Disparities Through AI*

The gap bridging ability of AI between the different socio-economic groups is one of the most significant contributions of AI in education (Khan et al., 2024). They enable us to help address educational disparity as follows.

- **Reducing the Cost of Quality Education**: Online courses are driven by AI and the grading is automated plus Operation cost for the education goes down, hence Quality education becomes more affordable. Because AI is capable of freeing

Table 1 AI-driven solutions for educational access

AI solution	Description	Impact on educational access
AI-powered learning platforms	Web-based platforms which employ AI for the customization of materials	Serves remote learners as well as those outside the traditional learning framework
AI-powered language translation	The automatic conversion of educational content into different languages	This helps erasing the language constraints in education
Assistive AI technologies	Voice recognition, text rendering, smart devices or other such technologies	To help learners who have physical difficulties in order to be inclusive
AI chatbots and tutors	Computer programs of which facilitate learning by simulating the interaction between instructor and student	Enhances availability of educational resources and makes one learn on their own
Intelligent tutoring systems	We provide customized programs that make use of Artificial Intelligence	Increases the process of students understanding and involving
AI-based learning analytics	Insights of data collected about an individual learner's progress and academic performance	Prompts required responses and solutions to learning difficulties

oneself from the physical infrastructure of a school and the associated personnel, it can provide access to quality education in a fraction of the usual cost.

- **Adaptive Learning Through AI**: AI based adaptive learning platforms deliver content as per the progress of a student so that it is tailored to the learning needs of a student individually. These are platforms that use the machine learning algorithms to evaluate student performance, and they provide dynamic instruction.
- **Early Intervention through Predictive Analytics**: Students at Risk are identified using predictive analytics and the interventions are delivered to the students at the right time to improve the learning outcomes. Using attendance patterns, assignment submission and test scores, AI can identify students that require help, and facilitate with educators to attend to the same before the students lag behind.
- **AI in Open Educational Resources (OERs)**: AI powered Open Educational Resources (OERs) make free and free materials available to students all over the globe and promotes equitible access to quality education.
- **AI enabled Grading and Feedback Systems**: AI powered grading systems are being used to make automatic online evaluations, and providing instant constructive feedback allowing the students to quickly review and learn feedback.

2.3 AI's Impact on Marginalized and Underserved Communities

The potential for uplift of marginalised communities that were left out from the opportunities of the past is real (Danaher et al., 2013). Some notable initiatives include:

Refugee Education with AI: AI driven platforms make education available for free or low costs to the displaced students so that they do not stop learning. AI assisted learning is also provided by organisations such as Kiron and Duolingo for Refugees, specially for refugee populations.
AI Based Skill Development Programs: Furthermore, AI is being used for training of underserved populations in digital and technical skills to boost their employability in the organisations. For instance, AI for Social Good, Google's workforce training program featuring AI, provides learners with the skills in AI and data science that are job relevant.
AI in Rural Education: In rural education, AI is used to deliver high quality education to rural students through AI powered mobile learning applications resulting to a narrowing urban urban rural education divide. For illustration, the Indian AI4Bharat initiative uses AI for localization of educational contents to rural learner in their native languages.
AI for Special Needs Education: From Microsoft's Seeing AI, AI based solution help students with visual impairment to read learning materials through voice, transferring visual data into audio description.

2.4 Challenges in AI Implementation for Education Equity

Although a wide variety of benefits for the educational access (Khawaja & Karimi, 2024) to equity is brought by AI, challenges still remain:

Data Privacy and Security Concerns: Collecting loads of student data from the AI education system makes the system privacy and data security concerned. Consequently, trust is vital in AI powered education systems and its guardians should ensure stringent data protection measures in place.
Bias in Data and Decision making by AI Models: Algorithmic bias and fairness is also a phenomenon where AI models contain biases from data or decision-making processes that can lead to reinforcing existing inequities. The process of creating ethical AI systems, which are fair and inclusive, is required.
Digital Divide and AI Access Disparities: Not all the students are fortunate to have access to AI-powered education tools as they do not have the connectivity Internet (digital divide), or lack digital devices, or they are behind financially on due to the prevalence of these tools. It will require concerted effort focused on the part of various governments, educational institutions and technology providers.

Teacher Training and AI Integration: Educators Should Be Trained Properly for the Integration of AI Tools Into Their Teaching Methodologies. The effective use of AI in education remains uncertain unless teachers have enough preparation.

2.5 Future Directions for AI in Equitable Education

Future initiatives should focus on the following in order to help maximize AI's ability to make a positive difference in creating an inclusive and equitable education system (Jianzheng & Xuwei, 2023).

Developing Ethical AI Guidelines for the use of AI in education: The establishment of policies to assure an ethical use of AI in education if it is to prevent biases in the AI and ensure fairness for students (Paschal & Melly, 2023).

Expanding AI Accessibility in Low-Income Regions: Governments and tech companies should team up and offer lowcost, AI-based education solutions to the deprived localities.

Enhancing AI-Powered Open Educational Resources: More investment on AI enhanced OER can increase creation of good quality educational content for students all around the world without limitations based on one's financial background (Varatharajan et al., 2024).

AI Enabled Teacher Support Systems: AI can be trained to alleviate teacher workload through the task automation in administrative work and allowing the teachers to concentrate on the engagement of students and individualized education.

Therefore, relying on AI responsibly, higher education institutions and policymakers may assist in the creation of a more equitable and inclusive educational one. But as with any other technology, it has to be carefully implemented in a way that takes into account ethical challenges related to ethical implementation of AI, data privacy, equitable use and access to AI technologies, etc., to ensure that the AI benefits all learners regardless of their background.

3 Personalization of Learning Through AI

Artificial Intelligence (AI) is integrated into the system of higher education to the point that we are able to create highly individualized learning experiences based on the needs of the student. With the ability to adapt to different kinds of learner through AI, we can have a more dynamic instruction, intelligent tutoring that provides each learner customized support and increase learning outcomes based on the dimensions of the curriculums. The use of AI enables educators to guarantee that the pupils are getting education commensurate to their preferred pace, style and learning tastes. As personalization is being driven by AI, it also improves engagement as well as the knowledge retention and promotes lifelong learning. It discusses AI how has been used to customize education, the benefits that arise from AI driven personalization of education, the different key applications and the associated challenges.

3.1 Adaptive Learning Technologies

AI is used to adjust educational content depending on the performance of actual students engaging with the technologies (Ghergulescu et al., 2021). These are systems that help think about how students engage learning materials and how the curriculum can be adapted dynamically to the needs of the students. Key adaptive learning technologies include:

AI Powered Learning Management Systems (SMS): Blackboard, Canvas, and Moodle tie together AI in order to keep track of the student's performance and suggest a customized approach to learning (Krishna Pasupuleti, 2024). The use of AI driven analytics in these systems allow them to analyse a student's engagement and suggest resources that students can follow to achieve efficiency in their learning.

AI-Driven Assessments: AI driven assessment tools provide an intelligent method of analysis of the student's response to detect the sharp's and improvement's that were made by the student, adjusting subsequent questions to reinforce understanding. This serves students by allowing them focus on areas they have issues with and hence master difficult concepts much better.

Real-Time Feedback Mechanisms: By virtue of being AI based these platforms provide instant feedback on assignments, quizzes and projects to students so that they can infuse any changes before the manual grading is done. It helps students to get clarification immediately as well as reinforces learning in real time.

Personalized Content Delivery: This ensures that AI personalizes instructional content based on each student's learning style and learning style preference through offering multiple formats like video; audio; interactive simulation; or text-based resource as students are able to learn in the manner that works best for their brain.

AI-Based Virtual Tutors: Virtual tutors like Carnegie Learning's MATHia or Squirrel AI are AI based tutoring platforms, which offer an interactive support to the students, helping them solve the problem and providing real time assistance (Jothikumar et al., 2025).

All of these adaptive Learning Technologies and Their Benefits are summarised in Table 2.

Table 2 AI-driven adaptive learning technologies and their benefits

AI technology	Description	Benefit to personalized learning
AI-powered LMS	Suggests personalized posts based on the work of the student and tracks the progress of the student	Enhances engagement and motivation
Intelligent tutoring systems	AI tutors provide individualized support	Improves comprehension and retention
AI-driven assessments	It customizes questions on your performance	Identifies your weak areas and strengthens them further
Predictive analytics	Takes the learning patterns for a tailored instruction	Enables proactive intervention
Personalized content delivery	Adjusts the content format depending on the learner's learning style	Improves accessibility and engagement
AI-based virtual tutors	Provides real-time interactive support	Enhances problem-solving skills

3.2 *AI for Personalized Student Support*

Furthermore, the scope of AI includes lending personal, student support, for example, providing learners the necessary support to achieve success (Feng, 2024). AI-driven tools include:

AI Chatbots for Academic Assistance: Many universities deploy AI chatbots like IBM Watson Assistant to offer 24.7 support for students and respond to students' questions about coursework, deadlines and administrative tasks (Guo & Li, 2024). They improve student engagement by providing immediate guidance and relieve the human advisors workload.

AI Learns from Student's History: The AI analyzes students learning history and suggests the right courses, suitable study materials, and future career paths. Specifically, AI can suggest areas in which a student will improve by assessing what they have done in the past coursework, and point the student towards the right resources for targeted learning.

AI and the Scientific Method: AI can be harnessed to aid in scientific research by utilizing the scientific method to gather data and facts to be stored in the cloud, which students will have access to. By doing so, students are ensured to get guided by experts at each step that are keeping in view their aspirations and changing industry trends.

AI for Mental Health and Well-being: Some of this mental health and well-being AI is focused on analyzing student stress levels when they are using learning platforms, providing them mindfulness exercises, or connecting them to related support services. Some universities like the University of Southern California

employ AI to check the well-being of students and provide individualized and bespoke mental health interventions.
AI Driven Student Progress Monitoring: Through AI, dashboards are created to monitor student progress and alerting educators when progress of any student becomes a challenge, in a timely manner (Hadyaoui & Cheniti-Belcadhi, 2024). The strength of these tools is that they assist faculty in designing how they provide support to each student based on that student's individual needs.

3.3 Enhancing Learning Outcomes with AI

The improved learning outcome is via customized educational experience for students based on their needs (Saraswat, 2024). Key impacts include:

- **Increased Engagement**: The Personalized learning increases student's engagement as it makes the content more relevant and interactive. Games like reward-based learning, interactive simulations among others are used by AI driven platforms to keep students entertained.
- **Better Retention of Knowledge**: AI driven learning paths have repetition of concepts on the areas where a student has weaknesses ensuring they are better retained. AI-guided revisions of challenging topics ensures that these are repeated, and that this repetition in turn results in much better long-term retention.
- **Small Achievement Gaps:** AI will identify students who struggle with the most and provide personalized approaches in an effort to fill any marked testing holes. With this method, students requiring additional help are not left behind for long and are given their help as soon as possible.
- **Scalability in Education**: in personalisation of learning, where specific personalised instruction is automated for large number of students. In other words, universities can deliver high quality, custom learning experiences to thousands of students while incurring relatively modest increases in the resources of faculty.
- **Encouragement of Lifelong Learning**: AI makes the content adapt of one's abilities thus reinforced lifelong learning beyond degree programmes. A lot of the organizations use AI to create upskilling and reskilling programs based on the industry demands.

3.4 Challenges and Ethical Considerations in AI-Driven Personalization

With so many benefits to AI driven personalized learning, these challenges of AI driven personalized learning should not be overlooked (Bell et al., 2024).

- **Data Privacy and Security**: The availability of huge amounts of student data is an issue of privacy and security, which are paramount for accurate and personalized

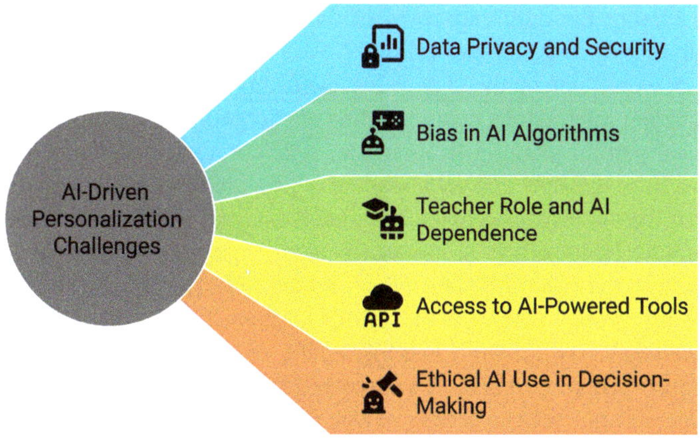

Fig. 2 Challenges and ethical considerations in AI-driven personalization

learning. For achieving trust in AI driven education, it is extremely important to ensure the compliance with regulations such as GDPR and FERPA.

- **Bias in AI Algorithms**: AI Algorithms can be Biased: From training data, AI models can learn the biases and apply it to training leading to unequal learning experience (Chaudhary, 2024). To serve algorithmic bias, our training dataset needs to be diverse, as well as transparent decision processes.
- **Teacher Role and AI Dependence**: How will Teacher Role and AI Dependence work? The over reliance in AI will leave the role of the educator minimal which will call for a more balanced approach. Human instruction should not be replaced by AI.
- **Access to AI-Powered Tools**: It is not affordable to all institutions to create independent organization of financial and technical resources for successfully implementing the use of AI driven personalization. All the AI driven solutions must be made accessible for all learners to bridge the digital divide and policymakers and education leaders have to work in this direction.
- **Ethical AI Use in Decision-Making**: Always ensure that the AI-based grading and feedback systems used are fair and free of any potential biases or any misinterpretations of student's performance.

All these challenges are being depicted in Fig. 2 as a smart diagram.

3.5 Future Prospects of AI in Personalized Learning

In artificial intelligence (AI), there is promising hope in revolutionizing personalized learning by transforming learning into the direction of individualized needs, learning styles and learning preferences of the particular student (Singh, 2025). Almost every fall, a new generation of college students starts bemoaning the death of the liberal education, cursing the dumbing down of the SAT, and decrying the rigidness of lecture courses they can only half follow.

The AI powered personalized learning platforms also take in large amounts of data from student interactions in order to find out areas of strength and weakness and the most ideal path of learning (Kumar Ale, 2023). It allows customized content to be delivered to make sure that each student is provided materials and task specifically created for his respective pace and understanding. In this way, these adaptive approaches have been proven to increase student engagement and help promote in learning through the upholding of motivation and decrease of frustration.

Additionally, AI powered tools like virtual tutor and intelligent assistants provide real time feedback in their work just like how a one-on-one mentorship work that used to be the privilege of a few. The democratization of personalized attention into educational environments can greatly reduce the gap in educational basis and lead to the creation of the more equitable learning environments. In addition, with the help of AI driven analytics, educators are able to get a glance at the student progress and take adequate measures where required.

For the future, whenever there is an ethical use, equity, and effectiveness of AI in personalized learning, stakeholders need to work together to create guidelines that guard against this. As technology becomes more advanced, AI will most probably become smarter at understanding the human cognition and provide more nuanced empathetic and supportive education. Therefore, with the integration of AI in education, the way of learning is likely to be completely transformed and the traditional classrooms may possibly be turned into dynamic and personalised environments for the cultivation of lifelong learners ready to face the complexities of the future.

4 AI in Administrative Efficiency and Institutional Management

A paramount role that Artificial Intelligence (AI) has taken in higher education administration is by acting as a powerful tool that has improved efficiency improved, simplified processes simplified and enhanced optimization of institutional management. The numbers attending universities and colleges are increasing, resources are scarce, and universities and colleges are forced to make good decisions. These challenges are addressed by AI technology through automating routine tasks, increasing decision support systems, and improving the experience of the student in general. This section discusses the role of AI in the administrative functions such as enrolment, grading, making decisions, allocating resources, and monitoring students thus helping in achieving Sustainable Development Goal 4 (SDG 4)—Quality Education.

4.1 AI in Enrolment, Grading, and Student Monitoring

Student enrolment is one of the costliest processes in higher education institutions. Admissions offices have traditionally been staffed by humans to apply, assess eligibility and admit applicants. This process has been revolutionized by AI powered systems, as they do many of these tasks (Khan et al., 2021). The machine learning algorithms can assess the student applications, review academic records, and commensurate student success from the historic data. It allows institutions to harmonize the admissions process and lessen the administrative costs while still maintaining a fair and fact-based approach for student selection.

We see an increase of AI powered chatbots and virtual assistants being utilized for enrolment in offering real time service to the prospective students. These resources assist applicants to understand the admission procedure, respond to frequently asked questions, and lead them in the way to their required paperwork (Hafiza et al., 2024). AI driven assistants enhance communication within an organization, resulting in better applicant experience, hence higher enrolment rates and more satisfied students.

Another area where AI is having a major impact in grading and assessment is the second most critical administrative area. The old grade assigning systems are not only time consuming, but are also likely to make human errors. This makes it possible for AI based grading systems to grade automatically assignments, quizzes, and even exams if the subjects have objective answers like mathematics, or sciences. Natural Language Processing (aka NLP) algorithms, too, improve the automated grading of essays, and of written responses in general. AI cannot completely substitute human evaluators when it comes to subjective assessment but it does a first landing of scoring and feedback, so that the instructors can spend more time on other important aspects like deep analysis and personalized feedback.

Student monitoring and academic performance tracking is also being revolutionized by people using AI. This is achieved by the predictive analytics models, which use AI to profile students most at risk of academic failure by identifying personalized behavioural patterns, attendance records, and most past performance of the students. Intervention strategies can be put in place at the earliest stages in order to offer extra assistance to struggling students in reducing dropout rates and enhancing educational outcomes in general. It sees through AI-powered monitoring systems in patterns of engagement and informs personalized learning intervention, and none of the students are left behind.

4.2 AI-Driven Decision Support Systems in Higher Education

Higher education decision making essentially boils down to managing a huge volume of performance data related to student outcomes, faculty quantities, financial planning and institutional efficiency. Decision support systems (DSS) are AI driven and the institutions learn actionable insights through data analysis that improve the institution's strategic planning and policy implementation (Zhang & Goyal, 2024). Based on data and machine learning, they help the administration make better decisions on administration for more efficient operation.

Student retention and success strategies is one of the main uses of the AI driven DSS. University data sources include attendance records, course performance scores and attendance levels. Such data is analyzed by AI models to detect if there are students needing support and to recommend good intervention methods (Lestari et al., 2024). For instance, quality academic advisors need to be alerted at the right time about the early risk of dropping out of a student that can facilitate proactive measure like counseling or tutoring support.

Artificial Intelligence helps us in financial management to make budget planning, allocate resource and so on (Gupta & Agarwal, 2024). Hence financial models that have AI driven will analyze expenditure pattern, predict the future financial need of a company and will help in an optimum distribution of resource. This help ensure institutions' efficient working, using limited resources and at the same time to produce high quality education. Moreover, AI based forecasting models guide in enrolment planning like institutions can anticipate the demand of the students and then accordingly allocate the resources.

Faculty and staff management also benefit from AI-driven decision support. AI can help compute faculty workload distribution, research and teaching productivity, and teaching effectiveness, which can help in optimizing the utilization of the human resource. This enables the institutions to make sure there is a balance workload of faculty members and pave way to excellence in teaching and research.

Institutional governance also integrates AI for policy making, which provides the decision makers evidence-based recommendations. A dashboard with the use of AI powered visualization tools and tools allows the administrator to make an informed decision by presenting complex data in an intuitive manner. These understandings help overall educational institutions to developing high quality education in the course of compliance with sustainable development guidelines. All the above points are depicted in Fig. 3 for convenience.

4.3 Automation and Resource Allocation

Better administrative efficiency in higher education institutions is achieved through automation. With an AI automation, there is a lessening of manual workload, lower incidence of human error in the processes and quickening of processes that would ordinarily take administrative staff some time to complete and, therefore, allow them to focus better on the strategic initiatives. Automation is available in various areas in higher education administration, such as student services, faculty support and infrastructure management.

Student services are one of the main uses of AI driven automation. Student inquiries, course registration and scheduling are handled by AI chatbots saving users

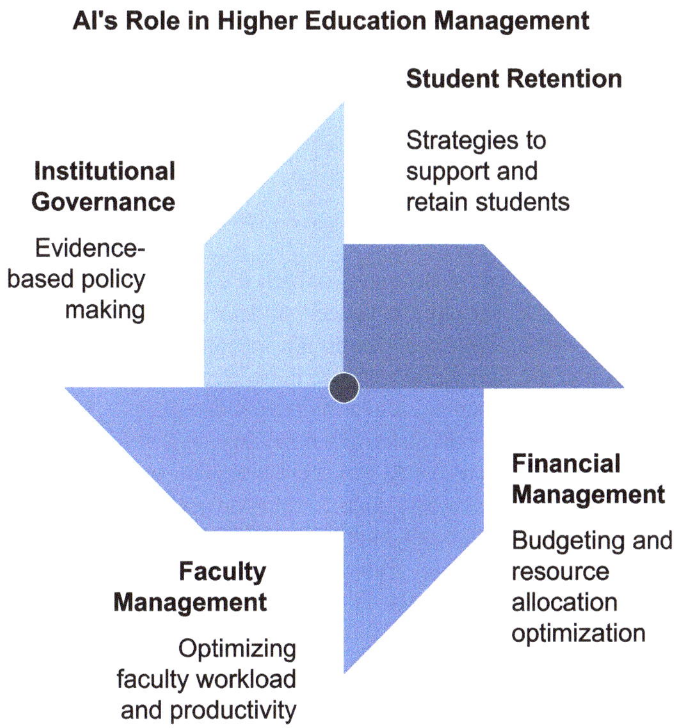

Fig. 3 AI's role in higher education management

time spent waiting and requires fewer human intervention to get accurate results. These chatbots run 24/7, giving students and faculty an opportunity to access such support at all times. The management of student records is also automated into the process through secure and efficient data processing.

Faculty support is automated through AI by eliminating roadblocks and remembering faculty's ongoings like grading, scheduling, and course management. Possible intelligent scheduling systems will make use of class timetable scheduling subject to faculty availability being respected, student preferences being addressed and respect for room capacity, finally utilising institutional resources efficiently. Moreover, AI helps in academic research through automation of data analysis, literature review, grant application, etc., which allows faculty to devote more time to innovative research tasks.

Another one of the most important areas where AI automated techniques help in improving the efficiency is the area of resource allocation. Classrooms, laboratories, and spaces for study must be allocated efficiently by the institutions to meet students needs. This worked in part because such AI powered systems allow historical iterations to be analyzed to predict future demand so that space can be properly allocated. These systems help utilize facilities effectively, thus saving operational costs and complementing good learning environment.

Moreover, AI driven smart campus solution provides infrastructure management. Energy consumption is monitored by AI enabled sensors and Internet of Things (IoT) devices, which also help in optimizing the heating and cooling systems inside the campus and also improving campus security. These are technos that help sustainably manage the campus, minimising wastage of energy and encouraging an ecofriendly practice.

They outline key areas in which AI driven automation has penetration in higher education administration in the Table 3.

The use of AI driven automation assists higher education institutions to have a more efficient operation, a more enhanced student experience and working for sustainable development objectives. Due to AI, institutions are freed from administrative burdens and are able to concentrate more on their areas of excellence by creating a lifelong learning and academic innovation campus.

Administrative efficiency and institutional management at higher education are being revolutionized with the use of AI which ensures a data led streamlined decision making. AI has impacted higher education administration from simple optimization of enrollment and grading through improvement of decision support systems to allocation of resources. AI Driven Automation offers the institutions a chance to reduce the operational costs, enhance the student services and provide the proper, sustainable and efficient learning environment.

Table 3 AI-driven automation in higher education administration

Area of administration	AI application	Benefits
Enrolment and admissions	AI chatbots, automated document verification	It is capable of faster application processing and leads to a better experience for students
Grading and assessment	NLP based essay evaluation, AI graded systems	Grading time can be reduced, students are graded in a consistent and fair way
Student monitoring	Predictive analytics for early intervention	Will result in improved retention rates, and with the help of a personal mentoring program, specific academic support
Decision-making	AI-driven decision support systems	Data-driven policies, strategic resource allocation
Faculty support	AI for workload management, research automation	It has better optimized faculty distribution, better research efficiency
Student services	AI chatbots, automated scheduling	24/7 support, streamlined administrative processes
Resource allocation	AI-powered space management	For instance, efficient use of campus facilities, cost reduction, etc.
Smart campus management	IoT-enabled AI systems	Energy efficiency, improved campus security

But the use of AI in administrative functions on a general level also brings with it ethical issues, such as data privacy, algorithmic racist, and availability of AI technology to all. Responsible AI governance frameworks must be implemented by institutions in order to guarantee that AI is used in decision making in a nonhumane and ethical way.

All in all, AI holds great promise for the improvement of higher education's quality, accessibility and sustainability in line with the general vision of SDG 4. With time, since its emergence AI will be a key player in administration of higher education and will set the platform for a future where AI driven institutions will operate efficiently and effectively.

5 Case Studies: AI-Driven Innovations in Higher Education for SDG 4

Bolstered by Artificial Intelligence (AI), traditional learning paradigms are being transformed in higher education, in many ways helping to address some of the key challenges that relate to Sustainable Development Goal 4 (SDG 4) on Quality Education (Sengupta, 2023). AI driven innovations are making for an inclusive and equitable education where all students can practically access it, learn better as per their unique needs and desires, and implementing greater institutional efficiency. Only using this section, I will provide a study of AI implementations in higher education in different global contexts using case studies, their impact, challenges and potentialities.

5.1 AI-Powered Learning Platforms

With this, the AI powered learning platforms have changed how students interact with educational content, making an experience of adaptive learning adapted to the need of an individual (Salman & Chaya, 2024). The machine learning algorithms of these platforms determine students' strengths and weaknesses, and make recommendations and real time feedback. This is one prominent example—Coursera's AI driven-course recommendations, for instance, analyses learner's behavior and provide them with courses relevant to their interests and what courses they have taken in the past. Likewise wise is Duolingo, a language learning platform based on AI, which personalizes exercises according to users' proficiency levels to help diversify learners to respond to a straight optimization of a learning curve.

There has been great success in personalized tutoring at the AI driven level by platforms such as Squirrel AI in China in higher education. Adaptive algorithms employed by Squirrel AI help the delivery of content to match the student's level, closing gaps in student understanding by finding the weak areas first and offering targeted exercises to strengthen it. Various studies have shown that, using Squirrel AI, students learn faster and comprehend more than what one can achieve with a traditional classroom. AI is shown to bring disparities in education down and to ensure education equity with these platforms.

5.2 Success Stories and Challenges

Although artificial intelligence provides a hope for a solution to the educational problems, its application is not fully problem free (Folgieri et al., 2024). The most successful case is the AI chatbot on Arizona State University (ASU) called Sunny, that supports students in handling administrative processes, accessing course materials and more academic support. Since ASU was deployed, the number of students engaged hours have gone up, administrative burden has decreased and retention rates have increased. Universities on a global scale are adopting similar chatbots so that students can get 24/7 support and students are assisted perpetually.

AI driven adaptive learning platforms has brought out another success story from India when used in the field of engineering education. Almost all of the institutions, like Amrita Vishwa Vidyapeetham, use AI techniques that are used to personalize the coursework so that every student can learn at his or her pace. This has helped to enhance student performance, personal learning experience, for instance, first generation college students who may lack academic support from home.

While these successes have not yet overcome the challenges of adoption of AI in higher education, these platforms would be more advantageous if the teachers felt interested in adopting them. The main barriers are algorithmic bias, inadequacy of the infrastructure, and unwillingness to switch to technological change. AI systems cope with the data of past events so it may harbor latent biases biasing the same to disadvantaged communities. AI models must be inclusive to prevent including the inequalities that already exist in education. Aside from this, constraints related to funding and technological access that institutions in developing nations experience, for instance, also stop the implementation of AI.

5.3 Future Potential of AI in Achieving SDG 4

AI in higher education will continue to change; emerging technologies are providing new ways of accomplishing SDG 4. AI enhanced Virtual Reality (VR) and Augmented Reality (AR) has the capacity to transform experiential learning and to better student's education experience. With AI driven voice recognition tools, presumably such tools will help students with disabilities as well so that educational resources can also become inclusive and equitable.

There is another promising development of predictive analytics for the improvement of students' retention. Through patterns found in a student's performance, AI can identify at risk learners and if timely intervention is offered the dropout rates can greatly decline and higher completion rates will be achieved for students' education. Furthermore, AI can enable lifelong learning in the sense that it can supply micro-credentialing and skill development programs that match up to rapidly changing job market in higher education.

In higher education, AI driven innovations are turning out to be transcendental in the advancement of SDG 4—guaranteeing unsurpassed and quality instruction for all. They include personalized learning platforms, AI powered institutional support system, etc. Despite challenges, proactive approaches in policymaking, ethical development of AI, and fair infrastructure can reduce risks and harness AI's potential for sustainable development. With that, higher education institution needs to wisely use these technologies as AI evolves to provide an inclusive, efficient and influential learning experience.

6 Ethical Considerations in AI for Higher Education

Artificial Intelligence (AI) in Higher Education has a great potential to revolutionize how education is accessed, learning experience is personalized and administrative efficiency is improved (Dangol, 2024). However, these benefits come at the cost of pressing ethical considerations that must be taken into account to realize any responsibility, equity, and fairness in AI deployment. Algorithms are biased, data is private, information is insecure, and AI tools are inaccessible to all. To ensure that AI is implemented in a manner that honours ethical principles, and that of Sustainable Development Goal 4 (SDG4) more broadly, it is essential to possess a solid framework of governance, transparency and accountability. In this section, I discuss the ethical pitfall for AI in higher education, and possible mitigation strategies maximize the potential benefits of AI in education.

6.1 Algorithmic Bias and Fairness in AI-Based Education

There is one of the major ethical concerns in higher education driven by AI and that is algorithmic bias. The decisions and recommendation made by AI models are based on large datasets. While these datasets can be skewed or partial, hence, the generated AI systems can enhance and propagate the same kind of biases (Jung, 2024). For an educational system, this can result in some youngsters, specifically, being disadvantaged as a consequence of gender, socioeconomic status, ethnicity, or geographical location.

For instance, a system for choosing admissions based on AI powered with historical data might possibly land up accepting candidates from privileged backgrounds

Table 4 Common sources of algorithmic bias and their educational implications

Source of bias	Description	Educational implications
Historical data bias	Since AI models learn from historical data, that data may contain systemic inequalities, and these can then be fed forward into the model	Past trends that have favoured Admissions AI towards privileged students
Sampling bias	Datasets are not representative of the full set of diversity in students	Underrepresented student groups have poorly performing AI tutors
Algorithmic bias	It so happens, that AI models magnify pre-existing biases in data	Automated grading that punishes non-standard style of writing
Implicit developer bias	Unconscious biases of developers appear in AI systems	Platforms learning did not take into consideration diverse learning styles of different cultures

without comprehending the issue itself. Just as AI based grading tools have the potential to bias against student whose style, dialect, and cultural background doesn't match that of the training data, if not designed with care.

Given that most AI decisions have the whiff of objectivity, fairness and scientific rigor, it can be difficult to query whether the results obtained are a consequence of sexist, racist, ethnocentric, ageist, or homophobic bias, so the problem of algorithmic bias is especially problematic. To resolve this problem, the training data used for AI models in education must be intentionally diverse, representative, and unbiased. Institutions also need to develop inspection and monitoring mechanisms that check for and rectify biases in AI's decision making on a regular basis.

The common sources of algorithmic bias and their educational implications are mentioned in Table 4.

The mitigation of bias in AI should include the use of explainable AI (XAI) models, transparent reporting of the AI decision making process, and engagement of the stakeholders for purposes of fair and inclusive deployment of the AI. Additionally, ethical AI training should be integrated into the curriculum developed for future educators and AI developers to raise awareness in such issues.

6.2 Data Privacy and Security Concerns in AI-Enabled Learning

AI is soon to be contributing an ever-bigger part in student assessment as well as in creating a tailored model for learning for each student, which is why it is imperative that there be a rightful collection and analysis of student data. In this setting data of the student: student data; are used by AI learning management systems, intelligent tutoring systems as well as predictive analytics tools to offer insights and recommendations (Jose Miss, 2024). However, a main cause for concern over these capabilities is that they enable collection of massive sets of personal data in such a considered and comprehensive manner.

The combination of student data for the benefit of personalized learning and prevention of misuse of student data and unauthorised access to student data is a challenge for educational institutions. Many such AI based learning platforms end up collecting data related to the students' level of progress, level of engagement, learning preferences and even behavioral patterns. However, if not handled in the proper manner, this data could lead to unintended outcome, like misused by the third party, data breach, or even surveillance issue.

One main problem in AI driven education is compliance of data protection laws like GDPR in Europe and FERPA in United States. Specifically, these regulations prescribe stringent data protection protocols whereby it is only permitted to collect, process the data of these students provided that they explicitly provide consent to the same. Additionally, institutions have to guarantee proper user privacy from the design of AI systems, with strong encryption methods and access controls enabled.

Compliance with regulatory policies is one of the fundamental reasons for having such a culture but it does not end here, as there are many more reasons. For instance, students and faculty members should be informed on their data being used, by whom, and how it is protected. Higher education can develop trust in AI and responsibly implement AI solutions in its procedures if the decision-making process can be transparent.

6.3 *Ensuring Equitable Access to AI Technologies*

However, if not equally distributed access for AI driven tools, AI can also widen the gaps for inequalities in education (Ajakaye & Pendergast, 2024). Imposing AI based learning systems on many institutions, especially in low-income areas is a big solicitation on their infrastructure, expertise or even financial resources to implement such systems. In addition, through a digital divide this can marginalise students from lower background who don't have availability to high quality AI driven resources.

A huge problem in guaranteeing that AI access is fair is that there is such a difference in the technological capabilities of different regions. Many students from the developing countries do not have high speed internet, modern computing devices, and much more due to AI enabled learning platforms. On a similar note, means of localized AI solutions catering to different linguistic and cultural contexts are also lacking for the purposes of AI driven education in regions that do not speak English.

Thus, policymakers and educational institutions should give importance to the inclusive AI deployment strategy to address the disparities of these areas. There is an opportunity for governments and international organisations to contribute in terms of funding AI-driven educational initiative that can reach underserved communities. Then, in addition to that, AI developers should concentrate on the development of multilingual, culturally adaptive AI systems that adapt to different needs of learning.

AI literacy programs can also be implemented in educational institutions to educate students as well as educators with the skills that will help them use AI effectively. Higher education institutions can implement their AI strategies alongside the targets of SDG 4 by promoting digital literacy and guaranteeing widespread access of AI-based learning resources.

6.4 Ethical AI Governance and Accountability

Given the ethical complexities with AI in education, meaning good strong governance frameworks need to be established to ensure responsible AI deployment in the first place (Lechterman, 2022). The development and use of AI in higher education should be governed by ethical governance which entails defining principles, guidelines and mechanisms of accountability guiding the development and use of AI. It must be ensured that AI technologies advance in an ethical manner with respect to human rights and promote educational equity in institutions.

The key components of ethical AI governance include,

- Awareness of the engineered process or function: AI models in education should be transparent, with the ability to be explained and interpreted to students and teachers alike, so as to understand how the AI was able to make a certain decision.
- Accountability: Institutions cannot avoid accountability for the outcomes of AI based decisions in case of AI driven decision making in place, students must have legitimate basis for challenging, or appealing, AI based assessment, and institutional integrity can be compromised without monitoring and counterchecking by students and teachers.
- Inclusivity: AI systems should have been built to include diverse thinking so as not to limit students from a certain background.
- Periodic Audits: AI systems should be audited from time to time assessing their fairness, accuracy and learning outcomes on people, especially those belonging at disadvantaged.

To create ethical AI policies in education, academia, government bodies and industry stakeholders have to collaborate. Instituting the discussion and research on ethical AI from an interdisciplinary perspective, institutions may assist in responsible AI development and deployment adhering to the principles stated in SDG 4.

6.5 A Path Toward Ethical AI in Higher Education

AI can definitely revolutionize the higher education with improving learning experiences, streamlining administrative processes, and giving access to good quality education. Whereas, ethical subject areas like algorithmic bias, data privacy, fair access to technological infrastructure, and AI governance sadly are lacking, which means that AI does not enhance education towards equity and inclusion (Rebholz et al., 2025). Ethical AI frameworks, unbiased AI models, transparent decision making, and reasonable access to AI enabled learning would allow institutions to take full advantage of the transformative power of AI, while also abiding with the principles of SDG 4.

With the rise of AI in education in higher education, ongoing research, involving stakeholders as well as oversight and regulation will continue to matter so that AI powered education is fair, inclusive, and supports the Sustainable Development Goals approach. A proper use of AI will require the institutions to be proactive in spreading the adoption of ethical AI and make sure that the technology is a tool for empowerment rather than exclusion. Addressing these ethical challenges will enable AI to be a catalyst of sustainable and quality education in the twenty-first century.

7 AI and Lifelong Learning for Sustainable Development

Artificial intelligence (AI) is rapidly changing every facet of education and especially for the purposes of lifelong learning (Said & Abdallah, 2024). With knowledge driven and technology dependent societies, education and skills development becomes not only a necessity but a necessity for the whole lifecycle of a person. To explain the importance of lifelong learning as the process of obtaining knowledge and skills during an individual's life, I emphasize that it is essential for person's growth, employability, and social inclusion. With AI integration into lifelong learning, people are accessing educational resources, using learning materials in new ways and developing skills in a new manner.

Personal and efficient educational experience is something that the future of AI powered learning solutions can bring to the education as a form of lifelong learning. By looking into what learners should look into in case they are to progress, what preferences they have, as well as what they intend their careers to be, AI can develop adaptive learning pathways that cater to personal needs. The use of the AI driven platforms makes the micro learning, virtual tutoring, and automated assessment possible thus making it easily accessible and engaging. These technologies relate to Sustainable Development Goal 4 (SDG 4) and aim at inclusive and equitable quality education and lifelong learning opportunities for all.

AI has the potential to revolutionize lifelong learning, but there are roadblocks to overcome: digital inequality, algorithmic bias, and the question of ethics. It is important to make the equitable access to AI-driven education the basis for the inclusive growth and sustainable development. In this section, we look at how AI is key to lifelong continuous learning, employee education, and overcoming challenges to lifelong education using AI.

7.1 The Role of AI in Continuous Learning and Upskilling

Artificial Intelligence (AI), is revolutionizing how we can learn continuously, develop our skills, and acquire a new knowledge throughout the entire life span. Technology and the labor market demand are going through rapid change, and in such an era, AI driven educational systems promise an era to individuals to participate in a lifelong learning to remain competitive and adaptable (Doherty et al., 2024). Dynamic customized pathways based on an individual's progress, skills and career aspirations is what AI makes possible for the world's learners to march towards Sustainable Development Goal target (SDG 4): Quality Education.

This includes formal education, vocational training and informal self directed learning and the way in which AI increases the affinity to lifelong learning is through adaptive learning models, intelligent tutoring systems and the automated content generation. The flexibility, accessibility offered by AI based platform including Massive Open Online Courses (MOOCs), AI powered virtual mentors, intelligent assessment tools etc. promote engagement of learners to a greater extent. These technologies enable people to learn at own pace, not dependent on geographical (like geography) and socioeconomic (like income) barriers, which are in line with the target under SDG 4 of ensuring inclusive and equitable quality education and learning opportunities throughout their lives.

Moreover, it can close the skills gap by suggesting certain courses based on the job market. AI makes learners receive allied educational content for career to improve employability and workforce ready. The learning and the feedback is also automated which helps learners to track their progress and find a place where they need improvement and therefore further support the culture of lifelong education.

One of the principal difficulties in lifelong learning is the digital gap, the impediment faced by groups of low endowment and oppressed regarding access to AI empowered education. In order to make sure that AI advances human development, governments and educational institutions need to work towards making AI powered lifelong learning initiatives available to everyone. The democratization of knowledge and knowledge as a factor for social mobility can be democratized by investing in open-access educational resources and digital infrastructure, so that AI becomes a tool for such a democratization.

7.2 AI-Powered Learning Models for Lifelong Education

AI functions as a system of systems with many learning models that enable lifelong education, involving such elements as personalised learning systems, AI backed microlearning, chatbots and virtual tutors, and automation of skill assessment with credentialing (Ahn, 2024). Personalized learning systems are built on the back of AI driven platform that use anonymized data to help design a customized learning

Table 5 AI-powered learning models and their impact on lifelong education

AI-powered learning model	Advantages	Challenges
Personalized learning systems	Has to follow custom learning pathways, engagement is high, everything is tailored on recommendations	Requires to collect extensive data and has to be sophisticated
AI-supported microlearning	Adaptable for the busy schedule, short and focused learning	May be shallow in a complex subject area
Chatbots and virtual tutors	Accessibility, 24-h support, and instant analysis of learner's behaviour	Limited capability to handle the nuances of the discussion regarding academic submissions
Automated skill assessment and credentialing	Recognition of objective evaluation and evaluation of gained skills	Risk of algorithmic bias in skill assessment

plan for the user. Such platforms guarantee learners are getting content that is appropriate for them and it is more effective in promoting engagement and retention. Microlearning that uses AI to help them pick up skills in short, modules at a time is a great way to learn for those with busy schedules; it's tightly packed bits of training applicable for the learner in their busy time.

Virtual tutors and chatbot can ensure accessibility by offering instant feedback on queries asked by the learner, assisting learner in solving complex problems. A self-paced learning which is more effective because these AI powered conversation agents improve engagement and comprehension. A way of automatically assessing skill and issuing credentialing in the form of a digital badge or certificate indicating what skills have been acquired. In this sense, this model is supposed to validate one's expertise in a structured and recognized way for the ones who look for career advancement. As shown in Table 5, this paper presents a comparative analysis of the AI powered Lifelong Learning models, and the advantages as well as challenges of these models with respect to each strategy.

Lifelong education is offered by AI driven learning models with a great potential, but accessing it, being equitable and reliable for all involved are necessary for sustained impact.

7.3 AI's Contribution to Sustainable Workforce Development

We cannot even think of workforce sustainability without lifelong learning, as it allows the professionals to re skill and upskill to suit the evolving needs of industry (Feltus, 2021). AI is helping fill the gap in between education and needs as it predicts future industry trends and demands that require skills. Labor market data are analyzed by AI algorithms, allowing the role of education to be adapted to the labor market by determining the emergence of emerging roles and the required skills with the training objectives proposed by educational institutions. This means students as well as professionals can ensure their learning with current and future workforce requirements.

In addition to this, AI helps to transition jobs as it offers career transition with the use of AI powered career advisory platforms that suggest courses and career pathways based on personal interest, prior experience and industry trends. By bringing together factors that enable online workers to transition into new roles or industries, these platforms provide targeted recommendations that enhance their employability. The AI powered training platforms for employees provides personalized learning experience and also benefited the corporate training and professional development. They assist employees to become industry ready by acquiring skills quickly and in a manner compatible with their work routine.

Automated skill assessment tools are another way AI uses to assess one's informal learning experiences and issue certifications. AI systems can assess the competencies of an individual and officially recognize the competencies they have gained unofficially in places other than the traditional academic environment. Especially in the case of people looking for jobs just for which they do not possess a formal degree, this seems to be a good idea to prove their expertise. AI is embedded into lifelong learning so that the future of work is also, adaptable, resilient, and resilient. By providing AI powered sustainable workforce development, they work in contact with SDG 8 (Decent Work and Economic Growth) as they are in line with job creation and skills acquisition, which result to knowledge-based economy.

7.4 *Overcoming Barriers to AI-Driven Lifelong Learning*

Although there is great promise in this domain of AI for lifelong learning, there are unfortunately some barriers to a true implementation of this at scale (Chapman et al., 2006). The digital divide which prevents several learners, particularly in developing areas to access AI based learning resources because of poor internet infrastructure and digital literacy is one of the most disruptive barriers. Closing this gap demands investment in digital infrastructure, as well as the creation of open-source AI learning platforms for all those of every location, and economical status.

Algorithmic bias and related ethical considerations are another critical challenge. Education typically displays some racial prejudice, and this prejudice can be expanded by the AI models as well. These biases force us to develop transparent AI because you need to disclose the biases within the dataset, and in addition, you need to encourage spread of more inclusive datasets while proposing the ethical frameworks to govern the use of AI in such a way that the same learning tool is fairer and accurate across different groups and people. A challenge comes from privacy and data security concern as AI powered learning system collects large user data to personalize the learning experience. To ensure data privacy there is need to ensure that users trust and therefore this can be enhanced through strong cybersecurity measures and as well regulatory frameworks.

Another issue facing the growth of AI models is also making them sustainable. Because the artificial intelligence (AI) systems require tremendous computational power, it raises concerns in relation to the energy consumption and in return detrimental to the environment. This can be alleviated through sustainable AI development that consists of energy efficient algorithms, eco-friendly computing practices that does not disperse any adverse environmental effects but maintains efficiency. To overcome these barriers, the solution should be developed as a multi stakeholder approach involving collaboration among the governments, the educational institutions and the technology providers to spur inclusive and sustainable AI driven lifelong learning solutions.

7.5 AI-Driven Lifelong Learning: A Pathway to Sustainable Development

Including AI for lifelong learning fits well with several of the additional United Nations Sustainable Development Goals (SDGs) above SDG 4. Through its way of functioning, AI contributes to economic growth (SDG 8), reduction of inequalities (SDG 10) and innovation (SDG 9). AI enables continuous learning to lay the basis for a sustainable society based on knowledge (Strielkowski et al., 2024).

For AI to fulfil its potential in lifelong learning, it will take greater effort to support the development of open-access AI learning platforms and make it easy for the learners from various parts of the world, and underserved communities, to benefit from it. Such platforms should be funded and supported by governments and international organizations so that all population groups can have access to the AI driven education. Institutions must also create ethical AI frameworks for educational AI applications so that fairness, transparency, and accountability precede such implementation. Education itself has to be integrated with learning AI literacy and digital skills to prepare the learners with necessary competencies to thrive in AI powered learning environments. Fostering the development of AI in lifelong learning will also serve to strengthen public private partnerships between governments, universities and the private sector in this focus.

By adopting these strategies, AI becomes a means for enabling lifelong learning and sustainable development and promoting global education in the most equitable and sustainable ways, creating resilient workforce.

8 Future Directions and Recommendations

The rise of Artificial Intelligence (AI) in higher education is advancing rapidly and offers tremendous opportunities on the improvement of learning, administration and research. Although, to effectively take advantage of AI's transformative potential, good strategies need to be defined to allow its use sustainably and equitably (Chiu, 2024). AI must serve not only educational excellence, but also work towards realising United Nations' Sustainable Development Goals (SDGs), especially SDG 4, Quality Education.

Despite this, there are still various challenges such as ethical issues with data, privacy of data, internet divide, and the possibility of over-reliance on technology instead of traditional approaches of pedagogy (Memarian & Doleck, 2023). Therefore, the future efforts would primarily be dedicated to creating AI driven innovations sensitive, ethical and adaptive to all the varying needs of learners globally. Secondly, there is the need for the development of policy frameworks for guiding the deployment of the AI in education, and the collaboration of governments, education institutions and technology developers.

It discusses the future opportunities for AI in higher education as this will define what the future of AI can bring in, policy recommendations, and how to bring these ideas to reality for SDG 4. If we take a deeper look at these three key areas of education, we can assure that AI will be an enabler for change in the right direction to ensure the reach of quality, accessible and lifelong learning opportunities to everyone.

8.1 Opportunities for AI-Driven Educational Innovation

By demonstrating such potential to revolutionize higher education in the accessibility, personalization, and administrative efficiency, artificial Intelligence (AI) has already taken a leap in proving its potential to take over the higher education. But its function for shaping the next form of education is much greater than current applications (Essien, 2023). In this visionary future, the use of AI in HE must emphasize innovation, inclusion and confrontation of new challenges towards achieving the sustainable development goals (SDGs), especially SDG 4: Quality Education. AI has the potential to reshape the traditional pedagogical models and will bring classroom learning according to student centric and outcome driven model.

The intelligent tutoring systems (ITS) (Ferster, 2022) are among the most promising opportunities. These systems can tailor themselves to different levels of understanding of students and to different learning styles, giving at the same time different kinds of feedback and guidance. In the future, with the ongoing development of AI, upcoming ITS will employ improved NLP models and deep learning techniques to make possible more natural, human behaviour-oriented interaction between the students and AI tutors. However, such systems can be used as supplemental learning sources, which will be effective in areas where there are not enough qualified teachers.

A vast application area of AI driven innovation forms the development of immersive learning environment. When AR and VR are teamed with AI, it is possible to create interactive trainings that provide students the ability to have hands on experience in medicine, engineering and even environmental sciences. Within these environments, AI can scan student engagement and adjust the difficulty or content given to the students as each learner progresses at the right pace.

Moreover, with AI, it's possible to democratize education with it, as we can create a multilingual learning platform that can be understood by numerous languages. AI aimed at language translation has already been used on many scales, and further developments in real time translation and contextual adaptation will remove language barriers, making quality education available to non-original speakers. AI can help instil cross cultural learning experience, which helps students from varied linguistic backgrounds come together and learn about the same curriculum without the need of the language limitations.

8.2 Policy and Frameworks for Sustainable AI in Education

Even though AI in higher education holds great potential, it is not commonly used today because there is no well-defined governance and policy in place (Chan, 2023). Policymakers, educational institutions and technology developers need to jointly work to establish ethical and regulatory guidelines of the use of AI in education to make sure AI is an enabler of sustainable development instead being a tool to amplify inequalities.

Integrating AI ethics into education systems is one of the main policy considerations. Ethical guidelines need to be developed by institutions regarding AI driven decisions and ensure that fairness, transparency, and accountability remains intact in all of it. This is especially important in admissions, grading and student assessment where biased algorithms could be used to enhance any existing inequalities. Algorithmic audits and fair assessment are, therefore, required to be mandated by regulatory bodies for identifying and mitigating biases in AI models.

Data privacy and security policy are also important to structure the future of AI in HTI. As AI platforms begin relying more and more on AI for analysis of their students' data, security o issues arise regarding consent and the misuse of personal information. Governments and institutions of education should impose tough data protection codicils that typically conform to universal configurations for example the General Data Protection Regulation (GDPR) to ensure security of student data and regulate use of AI responsibly. Moreover, institutions have to educate students and faculty members about the data privacy rights and ethical AI use so as to create awareness and responsibility.

Furthermore, AI policy frameworks should account for the digital divide when ensuring access to life enabling portions of the wider system of AI. To ensure no inequality in educational opportunities there is a need for governments to invest on the infrastructure including reliable internet access and AI compatible devices. In order to encourage AI adoption in poor regions, special incentive and funding programs should be offered to integrate students in marginalized communities to AI enhanced learning experiences.

Secondly, AI literacy must also be part of educational policies in such a way that students and educators are taught the essential skills to be able to work seamlessly with AI systems. Design of AI literacy programs should be developed to enable learners to learn about AI fundamentals, ethical issues, and practical usage. The understanding of how AI works in educational institutions helps the students make use of it with responsibility, and also helps them critically evaluate how it might work in different domains.

8.3 Pathways to Achieve SDG 4 Through AI

In applying AI in higher education, the second level of alignment is around how AI is integrated strategically into higher education and accordingly into the implementation of SDG 4: Quality Education (Gudonienė et al., 2021). This success should be attained after the use of AI since the intent is to make learning more inclusive, have better educational outcomes, and enable lifelong learning opportunities for everyone irrespective of their socioeconomic background.

Access to education in remote and underserved areas is one of the key ways to enable SDG 4 using AI. With the use of AI powered mobile learning applications and online platforms there are plenty of high-quality educational resources that can be provided to the students who have limited or no access to the traditional classroom settings (Trikha, 2024). Tailoring learning material to the specific needs of students, content can be personalized so that the student receiving the learning is receiving it based on their pace of learning and their attributes and way of learning.

In terms of SDG 4, a crucial role of AI is to promote lifelong learning and upskilling initiatives. With the pace of advancements in technology and the change in job market requirements, perpetual learning has become inadequate for advancement in your career. AI can help lifelong learning by providing the individual with the perfect learning pathways, tailored to the competencies that they have or the career goals they have set out to achieve. There are AI driven recommendation systems that will suggest suitable courses, certifications and skill development programs that can help the person to stay up to date in his/her field.

AI can also contribute much in the improvement of teacher training and professional development. Educators can use AI powered analytics to understand student learning pattern and how to change their teaching style in this process to support students who need more help. With AI driven teacher training platforms, you can get personalized suggestions for professional development programmes for the teachers

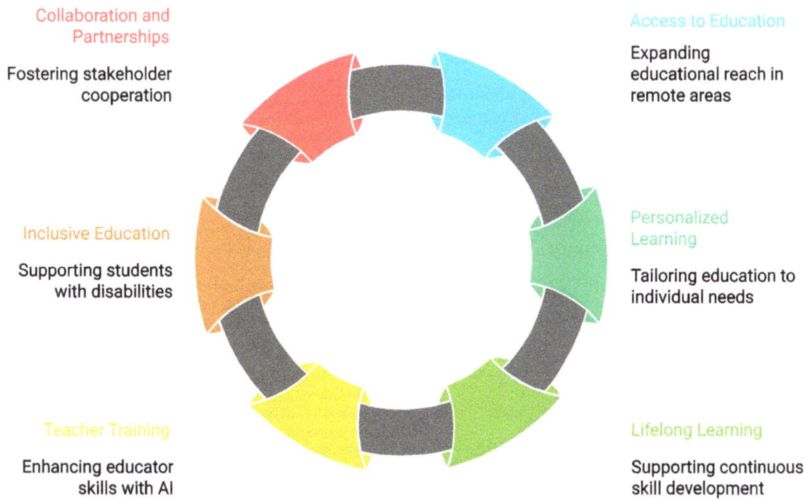

Fig. 4 Ways to use AI to reach SDG 4

so that they can be up to date with the latest teaching methods and technology taught in the curriculum.

Moreover, AI can be helpful towards an inclusive education by support for students with disabilities. Narrators Explain further how AI technologies such as Speech to Text, Screen Readers and AI Sign Language Translation Tools reduce the barrier of disabilities for the students in seeing and benefiting from the educational content in the same way as their counterparts. With the aid of AI, institutions can design environments that give learning more inclusivity to help more diverse needs and make institutions equally open to all learners.

Achieving SDG 4 will require collaboration between governments, academia and the private sector, and AI technology has the potential to play such a role. Partnerships with other stakeholders involved in the promotion of AI in education can also enable knowledge sharing, technology advancement and mobilization of resources for driving the AI powered educational initiatives. However, funding gap may hinder the implementation of such AI driven project of sustainable development in education, for which public private partnerships (PPPs) can help to bridge the gap. The ways to Achieve Sustainable Development Goal 4 with Artificial Intelligence are depicted in Fig. 4.

9 Conclusion

The future of higher education is being largely redefined by Artificial Intelligence (AI) and is transforming the United Nations' Sustainable Development Goal—Quality Education (SDG 4). Thanks to the features of AI, the availability of learning, personalization of education and simplification of the processes of administration of education become functions of a powerful tool of sustainable educational transformation. While we headed towards the future, where modern digital innovation plays an important part of the educational systems around the world, we need to analyze both, opportunities and challenges, AI can provide.

AI can be one of the vital contributions in the development of higher education as it can bridge the gap of educational challenges and allows inclusive and consistent access to quality education. There is already a heavy impact that AI made by AI powered tools like intelligent tutoring system, automated grading and adaptive learning platform. They offer personalized paths of learning to students, and education will be suited to their needs, their learning paces, their preferences. For example, these self-learning bots and virtual assistants with AI powers allow 24/7 academic help and reduce the need of teacher led instruction for learning as they help people learn on their own. Moreover, AI can use large scale educational data to identify which students are at risk of not surviving, and suggest early intervention strategies to keep the students from leaving, increasing student's retention rates and overall educational success.

In addition to enhancing the learning experience of individuals, AI plays a crucial role in optimizing institutional operation and administration in higher education. Nowadays universities and colleges are taking up the use of AI driven tools for student enrollment, course allocating and performance analysis. Educational institutions make use of AI powered analytics systems to track student progress, to identify patterns in students' learning behavior and help them with data driven insight which administrators and educators can make use of. It allows for the improvement in decision making processes which consequently guarantees effective use of resources in the universities to improve student learning outcomes. Additionally, the automation of administrative tasks with AI relieves the faculty and staff from performing lower value tasks so they can focus more on conducting teaching and research activities.

With regard to higher education, an important role of AI is developing lifelong learning and skills development, which are fundamental to sustainable development. In times when technology is developing at the speed of light and the labour market continues to change, AI driven learning platforms offer learners an opportunity to continue gathering new skills and new knowledge. The use of AI is for building flexible, competency driven, learning methodologies that allow individuals to continue to remain relevant as the workplace becomes more digitized. The usage of Massive Open Online Courses (MOOCs) and AI powered training programs has enabled people from all kinds of diverse populations like the working professionals, the underserved communities and people residing in remote areas have access to learning. With AI, lifelong learning also becomes a more dynamic, more adaptable and more inclusive process, consistent to the aim of SDG 4 to provide quality education to all.

However, integrating AI in higher education also entails several ethical and socioeconomic challenges that need to be tackled to enable the appropriate adoption of this technology in higher education. As one of the key concerns, algorithmic bias can unveil the discrimination issue and amplify educational inequality. When AI is trained on the biased data, the outcomes are also biased, which may disadvantage certain groups of student's disproportionalities. This requires institutions to adopt the ethical AI principles and thus, the necessary data governance framework that fosters fairness and transparency during the AI decision making process.

References

Ahn, H. Y. (2024). AI-powered E-learning for lifelong learners: Impact on performance and knowledge application. *Sustainability, 16*(20), 9066. https://doi.org/10.3390/su16209066

Ajakaye, J., & Pendergast, N. (2024, May) AI generated metadata for equitable publication access. In *AI generated metadata for equitable publication access*. ScienceOpen. Retrieved April 02, 2025, from https://doi.org/10.14293/s2199-ssp-am24-01025

Bell, C., Olukemi, A., & Broklyn, P. (2024, August). *AI-driven personalization in digital marketing: Effectiveness and ethical considerations*. MDPI AG. Retrieved April 02, 2025, from https://doi.org/10.20944/preprints202408.0023.v1

Chan, C. K. Y. (2023). A comprehensive AI policy education framework for university teaching and learning. *International Journal of Educational Technology in Higher Education, 20*(1) https://doi.org/10.1186/s41239-023-00408-3

Chapman, J., McGilp, J., Cartwright, P., de Souza, M., & Toomey, R. (2006). Overcoming Barriers that Impede Participation in Lifelong Learning. In *Lifelong learning, participation and equity* (pp. 151–174). Springer Netherlands. Retrieved April 02, 2025, from https://doi.org/10.1007/1-4020-5322-3_8

Chaudhary, S. (2024). Cognitive bias in AI: Identifying and mitigating human-like flaws in algorithms. Elsevier BV. Retrieved April 02, 2025, from https://doi.org/10.2139/ssrn.4971324

Chiu, T. K. F. (2024). Future research recommendations for transforming higher education with generative AI. *Computers and Education: Artificial Intelligence, 6*, Article 100197. https://doi.org/10.1016/j.caeai.2023.100197

Danaher, M., Cook, J., Danaher, G., Coombes, P., & Danaher, P. A. (2013). Situating education research with marginalized communities. In *Researching education with marginalized communities* (pp. 1–22). Palgrave Macmillan. Retrieved April 02, 2025, from https://doi.org/10.1057/9781137012685_1

Dangol, D. L. (2024). Ethical considerations and pedagogical challenges of integrating AI in higher education in the UK. *Welcome to the International Journal Multidisciplinary Business Management, 12*(5), 23–29. https://doi.org/10.56805/ijmbm.2024.12.5.105

Darıcan, Ş. (2025). Artificial intelligence in education and its importance. In *Advances in educational technologies and instructional design* (pp. 91–116). IGI Global. Retrieved April 02, 2025, from https://doi.org/10.4018/979-8-3693-7949-3.ch004

Dawson, S. (2021, March). *SDG 4: Quality education*. https://doi.org/10.14293/s2199-1006.1.sor-socsci.cl6f5fl.v1

Doherty, I., Huijser, H., Willems, J. (2024, November). *Upskilling academics for Gen AI* (pp. 658–663). ASCILITE Publications. https://doi.org/10.14742/apubs.2024.1509

Essien, A. (2023). AI-driven innovation: Leveraging big data analytics for innovation. In *Innovation analytics* (pp. 119–137). World Scientific (Europe). Retrieved April 02, 2025, from https://doi.org/10.1142/9781786349989_0006

Fathima Shah, W. (2024). Enhancing healthcare access and equity: The role of technology and digital solutions in CMSs strategic partnerships. *International Journal of Science and Research (IJSR), 13*(3), 829–832. https://doi.org/10.21275/sr24311180631

Feltus, C. (2021). AI'S contribution to ubiquitous systems and pervasive networks security—Reinforcement learning vs recurrent networks. *Journal of Ubiquitous Systems and Pervasive Networks, 15*(02), 01–09. https://doi.org/10.5383/juspn.15.02.001

Feng, M. (2024). Support student math learning with a personalized learning system that responds to student affect (Poster 38). In *AERA 2024*. AERA. Retrieved April 02, 2025, from https://doi.org/10.3102/ip.24.2103474

Ferster, B. (2022). Intelligent tutoring systems. In *Intelligent tutoring systems*. Routledge. Retrieved April 02, 2025, from https://doi.org/10.4324/9781138609877-ree6-1

Folgieri, R., Gil, M., Bait, M., & Lucchiari, C. (2024). AI-powered personalised learning platforms for EFL learning: Preliminary results. In *Proceedings of the 16th international conference on computer supported education* (pp. 255–261). SCITEPRESS—Science and Technology Publications. Retrieved April 02, 2025, from https://doi.org/10.5220/0012672000003693

Ghergulescu, I., Flynn, C., O'Sullivan, C., van Heck, I., & Slob, M. (2021). A conceptual framework for extending domain model of AI-enabled adaptive learning with sub-skills modelling. In *Proceedings of the 13th international conference on computer supported education* (pp. 116–123). SCITEPRESS—Science and Technology Publications. Retrieved April 02, 2025, from https://doi.org/10.5220/0010451201160123

Gudonienė, D., Paulauskaitė-Tarasevičienė, A., Daunorienė, A., & Sukackė, V. (2021). A case study on emerging learning pathways in SDG-focused engineering studies through applying CBL. *Sustainability, 13*(15), 8495. https://doi.org/10.3390/su13158495

Guo, K., & Li, D. (2024). Understanding EFL students' use of self-made AI chatbots as personalized writing assistance tools: A mixed methods study. *System, 124*, Article 103362. https://doi.org/10.1016/j.system.2024.103362

Gupta, A., & Agarwal, P. (2024, May) Enhancing sales forecasting accuracy through integrated enterprise resource planning and customer relationship management using artificial intelligence. In *2024 3rd International conference on artificial intelligence for internet of things (AIIoT)* (pp. 1–6). IEEE. Retrieved April 02, 2025, from https://doi.org/10.1109/aiiot58432.2024.10574785

Hadyaoui, A., & Cheniti-Belcadhi, L. (2024). IntelliFrame: A framework for AI-driven, adaptive, and process-oriented student assessments. In *Proceedings of the 20th international conference on web information systems and technologies* (pp. 441–448). SCITEPRESS—Science and Technology Publications. Retrieved April 02, 2025, from https://doi.org/10.5220/0013070800003825

Haron, N. H., Mahmood, R., Amin, N. M., Ahmad, A., & Jantan, S. R. (2024). An artificial intelligence approach to monitor and predict student academic performance. *Journal of Advanced Research in Applied Sciences and Engineering Technology, 44*(1), 105–119. https://doi.org/10.37934/araset.44.1.105119.

Helium niposium. (2023, May). *The role of AI in sustainable development: Challenges and prospects*. Center for Open Science. Retrieved April 02, 2025, from https://doi.org/10.31219/osf.io/64pfh

Jianzheng, S., & Xuwei, Z. (2023). Integration of AI with higher education innovation: Reforming future educational directions. *International Journal of Science and Research (IJSR), 12*(10), 1727–1731. https://doi.org/10.21275/sr231023183401

Jose Miss, D. (2024). Data privacy and security concerns in AI-integrated educational platforms. *Recent trends in Management and Commerce, 5*(2), 87–91. https://doi.org/10.46632/rmc/5/2/19

Jothikumar, K., Anisha, S., Aruna Varshini, S., Meenakshi, A. J., & Gomathi, V. (2025). AI tutors and virtual classrooms. In *Driving quality education through AI and data science* (pp. 503–530). IGI Global. Retrieved April 02, 2025, from https://doi.org/10.4018/979-8-3693-8292-9.ch022

Jung, W. (2024). 'AI for humanity': Fairness and the optimization of bias in AI. *The Journal of Human Studies, 64*, 5–15. https://doi.org/10.33638/jhs.64.1

Kaushik, M., Sharma, M., & Choudhary, M. (2024). Sustainable management with AI: A study of the role of artificial intelligence in achieving the sustainable development goals. In *Multidisciplinary approaches for sustainable development* (pp. 124–129). CRC Press. Retrieved April 02, 2025, from https://doi.org/10.1201/9781003543633-20

Khan, I., Ahmad, A. R., Jabeur, N., & Mahdi, M. N. (2021). An artificial intelligence approach to monitor student performance and devise preventive measures. *Smart Learning Environments, 8*(1). https://doi.org/10.1186/s40561-021-00161-y

Khan, S., Rathod, V., Ranjan, A., Una, A., & Pandya, N. (2024). Addressing educational disparities: Assessing the gap for indigenous community. In *Proceedings of the 16th international conference on computer supported education* (pp. 453–460). SCITEPRESS—Science and Technology Publications. Retrieved April 02, 2025, from https://doi.org/10.5220/0012253600003693

Khawaja, S., & Karimi, H. (2024). AI in English higher education: Balancing innovation with equity challenges and opportunities. In *An overview of literature, language and education research* (Vol. 3, pp. 55–66). B P International. Retrieved April 02, 2025, from https://doi.org/10.9734/bpi/aoler/v3/1147

Krishna Pasupuleti, M. (2024). Green energy revolution: AI-powered systems for sustainable power management. In *Empowered future: AI-driven green energy systems for sustainable power management* (pp. 67–81). National Education Services. Retrieved April 02, 2025, from https://doi.org/10.62311/nesx/7264

Kumar Ale, N. (2023). AI and machine learning in test automation: Techniques, challenges, and future prospects. *International Journal of Science and Research (IJSR), 12*(12), 2130–2131. https://doi.org/10.21275/sr24717103320.

Lechterman, T. M. (2022). The concept of accountability in AI ethics and governance. In *The oxford handbook of AI governance* (pp. 164–182). Oxford University Press. Retrieved April 02, 2025, from https://doi.org/10.1093/oxfordhb/9780197579329.013.10

Lestari, R., Windarwati, H. D., Hidayah, R. (2024). AI-driven decision-making applications in higher education. In *Advances in media, entertainment, and the arts* (pp. 246–268). IGI Global. Retrieved April 02, 2025, from https://doi.org/10.4018/979-8-3693-0639-0.ch011

Lowa, M. (2024). Democratizing access to education: AI-driven solutions for inclusive learning in Malaysia. *SSRN Electronic Journal*. https://doi.org/10.2139/ssrn.4813758

Mandal, J. (2020). AI integration in lifelong learning: Transformative approaches to teaching and learning. In *Artificial intelligence in education: Revolutionizing learning and teaching*. Red Unicorn Publishing. Retrieved April 02, 2025, from https://doi.org/10.25215/9358094575.27

Memarian, B., & Doleck, T. (2023). Fairness, accountability, transparency, and ethics (FATE) in artificial intelligence (AI) and higher education: A systematic review. *Computers and Education: Artificial Intelligence, 5*, Article 100152. https://doi.org/10.1016/j.caeai.2023.100152

Paschal, M. J., & Melly, I. K. (2023). Ethical guidelines on the use of AI in education. In *Advances in educational technologies and instructional design* (pp. 230–245). IGI Global. Retrieved April 02, 2025, from https://doi.org/10.4018/979-8-3693-0205-7.ch013

Rebholz, S., Libbrecht, P., & Müller, W. (2025). Designing trustworthy AI in higher education. In *AI—Ethical and legal challenges [Working Title]*. IntechOpen. Retrieved April 02, 2025, from https://doi.org/10.5772/intechopen.1009134

Salman, S., & Chaya, R. (2024). The influence of AI-powered learning platforms on student engagement and performance: Emerging technologies in education. *International Journal of Research Publication and Reviews, 5*(7), 1816–1824. https://doi.org/10.55248/gengpi.5.0724.1750

Said, S. A., & Abdallah, W. (2024). Lifelong learning for sustainable development aligning adult education with sustainable development goal 4. In *Studies in systems, decision and control* (pp. 259–269). Springer Nature Switzerland. Retrieved April 02, 2025, from https://doi.org/10.1007/978-3-031-62106-2_21

Saraswat, D. (2024). AI—Driven pedagogies: Enhancing student engagement and learning outcomes in higher education. *International Journal of Science and Research (IJSR), 13*(11), 1152–1154. https://doi.org/10.21275/sr241119221041

Sengupta, E. (2023). Introduction to higher education in emergencies: International case studies. In *Innovations in higher education teaching and learning* (pp. 3–10). Emerald Publishing Limited. Retrieved April 02, 2025, from https://doi.org/10.1108/s2055-364120230000052001

Singh, A. (2025). The future of learning: AI-driven personalized education. *SSRN Electronic Journal*. https://doi.org/10.2139/ssrn.5076438

Strielkowski, W., Grebennikova, V., Lisovskiy, A., Rakhimova, G., & Vasileva, T. (2024). AI-driven adaptive learning for sustainable educational transformation. *Sustainable Development*. https://doi.org/10.1002/sd.3221

Trikha, A. (2024). Towards implementing SDG-4: Leveraging AI in today's educational landscape (pp. 69–79). In *World sustainability series*. Springer Nature Switzerland. Retrieved April 02, 2025, from https://doi.org/10.1007/978-3-031-68427-2_4

Varatharajan, A., Kumar, M. A., Sanchita, & Sandhya. (2024). AI-powered mobility educational application for enhancing students learning. In *AIP conference proceedings* (p. 020005). AIP Publishing. Retrieved 02, 2025, from https://doi.org/10.1063/5.0183574

Villarino, R. T. (2024, August). *Artificial intelligence (AI) integration in rural Philippine higher education: Perspectives, challenges, and ethical considerations*. Center for Open Science. Retrieved April 02, 2025. [Online]. https://doi.org/10.31219/osf.io/ehcb9

Yadav, S. (2024). Artificial intelligence (AI) integration in higher education. In *Impact of artificial intelligence on society* (pp. 43–59). Chapman and Hall/CRC. Retrieved April 02, 2025, from https://doi.org/10.1201/9781032644509-4

Zhang, J., & Goyal, S. B. (2024). AI-driven decision support system innovations to empower higher education administration. *Journal of Computers, Mechanical and Management, 3*(2), 35–41. https://doi.org/10.57159/gadl.jcmm.3.2.24070

Charting the Path to 2030 By AI in Higher Education

Aftab Ara and Swetapadma Das

Abstract This study investigates the effect of Artificial Intelligence (AI) in Higher Education to advance Sustainable Development Goal 4 (SDG 4), for an inclusive, equitable, and quality education for all by 2030. The study discusses the capabilities of AI to provide personalized learning experiences, opportunities for better engagement, more streamlined administration, and lifelong learning that can increase access and improve quality of education that are required in the world. However, AI brings significant challenges for education systems-ethical challenges, algorithmic bias, data privacy, disparities in educational infrastructure, and the digital divide. To address these challenges, this study shares a strategic roadmap, made up of short, medium, and long-term goals. The phased roadmap highlights the need for governance frameworks, ethical principles, and stakeholder engagement to facilitate the responsible use of AI. The roadmap entails international cooperation and strategic policies to close gaps in infrastructure and access. This study's key finding is that successful and responsible deployment of AI in education can be reached by a holistic approach using a core foundation of ethics, transparency, and accountability that can help build trust for stakeholders. This study underscores need to work strategically to consider the interaction of the existing pedagogical, technological, and policy systems with the role of AI, and create strong interdisciplinary paths to encourage and stimulate the transformation of higher education to more accessible, inclusive, and innovative educational system that aligns with SDG 4. It is the goal of this study for a coordinated effort between policy makers, educators and technology developers and providers for a sustainable education system in the world.

Keywords Artificial intelligence · Higher education · Sustainable development goal 4 (SDG 4) · Ethical challenges · Digital divide · Policy framework · Strategic roadmap · Responsible AI · Global collaboration

A. Ara (✉)
College of Business Administration, University of Hail, Hail, Saudi Arabia
e-mail: ara.aftab@gmail.com

S. Das
ASTHA School of Management, Bhubaneshwar, India

A. Ara (ed.), *Role of Artificial Intelligence in Higher Education*,
https://doi.org/10.1007/978-981-95-5350-1_8

1 Introduction

Artificial Intelligence (AI) in higher education can improve teaching, learning, and administration, by aligning with the Sustainable Development Goal 4 (SDG 4) ensuring inclusive and equitable quality education for all. AI tools using algorithms are used for personalization, enhancing efficiencies, and extending educational opportunities; innovation and sustainability would flourish, while improving the role of education for societal change. Yet, there are several challenges and gaps for the responsible utilization of AI in higher education. One challenge is the forms of risks after AI is implemented, including algorithmic biases, data privacy issues, infrastructural issues, and the digital divide; these issues have the potential to further our inequity, increase risk of bias, and undermine our ethical assumptions for educational practice. There are also fewer developed global staged AI government frameworks and stakeholder engagement procedures which could be a significant step in realizing equitable AI integration.

Most current research and policy focused on AI adaptation do not think about staged implementation of government planning, nor the vision or applications over the short-, medium-, and long-term. There is also a lack of practical definitions of governance, ethically responsible deployment/integration, frameworks for inclusive governance, and collaborative international approaches to address infrastructure and socio-economic disparity. Furthermore, more empirical data is warranted to support claims of AI related risk mitigation strategies that might shaped responsible policy and practice.

Objectives

1. Analyze opportunities and challenges linked with AI in higher education.
2. Identify the gaps in policy, governance and practice.
3. Propose a roadmap along with stakeholder-specific strategies to guide responsible AI adoption in alignment with SDG 4.

Implications

The study has implications for policymakers, educational institutions, technology developers, and other stakeholders to create a strategic and systematic approach for the responsible AI use for fair, equitable and quality education globally. This study aims for the development of resilient, innovative, and inclusive higher education systems that harness the potential of AI to achieve SDG 4 by 2030.

2 Literature Review

2.1 AI Applications for Enhancing Quality Education: SDG 4.1, 4.4 and 4.6

The integration of Artificial Intelligence (AI) into education presents many opportunities to improve the quality and accessibility of learning experiences across different educational settings. Following are the different aspects of how AI applications are transforming education as derived from the literature review:

1. **Personalized Learning and Instruction**: AI facilitates the personalization of learning experiences, enhancing the quality of education. Intelligent Tutoring Systems and adaptive learning platforms utilise AI to customise instruction based on individual learning needs of each student. These systems provide real-time feedback and adjust learning processes to optimize the educational outcomes (Lin et al., 2023). For instance, The Teaching Machine System uses predefined rules and AI algorithms to deliver personalized instruction (Lin et al., 2023).
2. **Enhanced Student Engagement**: AI tools, like chatbots and Generative AIs, improve student engagement with more interactive and engaging learning. These technologies can give feedback, engage students, and facilitate peer learning. Thus it gives students a more supporting learning experience.
3. **Improved Assessment**: With the help of AI tools teachers play an important role by catering to individual students learning stylesthus proving to be more accurate and efficient during evaluation (Owan et al., 2023). So as the teachers can focus more on instructional design and support of students as these tools automate the assessment. (Owan et al., 2023).
4. **Resource distribution with Equity in Education**: AI tools can gauge student's performance and help in resource allocation according to priority so that there can be an equal distribution of quality education (Alsagri & Sohail, 2024).
5. **Improving Teaching Effectiveness**: AI helps educators by helping in preparation of teaching materials, designing presentations, and providing accurate assessments (Nasir et al., 2024). The use of AI in education helps teachers with new methodology for teaching for diverse learners to improve student outcomes (Nasir et al., 2024).
6. **Lifelong Learning and Sustainability**: AI helps in lifelong learning thus aligning with sustainability goals to improve equity in education and prepare students to face future challenges (Alsagri & Sohail, 2024).

These challenges can be addressed by collaboration among educators, policymakers, and AI developers to create effective strategies for AI's benefits and mitigating potential risks (Akinwalere & Ivanov, 2022; Owan et al., 2023). Further research and development in AI's role in education is critical in realizing its full potential (Alsagri & Sohail, 2024).

2.2 *AI for Equitable Access and Inclusion in Higher Education*

Aligning with Universal Design for Learning (UDL) principles AI can act as tutors that understand the students' pace and style. (Kohnke & Zaugg, 2025). AI is also quite helpful in developing accessible educational softwares that supports personalized learning paths and ensuring learning materials are appropriate for all students. The 'EQUAL AI'framework helps AI to improve equity, quality, and inclusivity in education. It leads the way for responsive policies to address challenges like algorithmic bias and the digital divide (Davoodi, 2024).

Ethical considerations play a crucial role, as responsible use of AI requires transparency and fairness, the absence of which can exacerbate disparities instead of mitigating them (Aprianto et al., 2024).

AI tools improve educational equity by reducing the digital divide and giving an inclusive learning environment (Edeni et al., 2024). To receive these benefit institutions require frameworks and policies for equitable access to AI technologies talking care that data privacy and ethical considerations are important (Aprianto et al., 2024).

2.3 *Ethical Considerations and Challenges of AI in Higher Education*

1. **Bias and Fairness in AI usage**: Diverse data training is required for training AI models by adhering to ethical guidelines (Barnes & Hutson, 2024).
2. **Data Privacy and Security**: By following the ethical guidelines student's personal information are more secure by preventing unauthorized use (Eden et al., 2024).
3. **Digital Divide and Accessibility**: To prevent digital divide among students there should be equal distribution of AI technologies in different socio-economic groups is ensured (Aprianto et al., 2024).
4. **Academic Integrity**: Increase use of AI for automating tasks reduces the critical thinking of students and educators, these issues need to be balanced (Khatri & Karki, 2023).
5. **Ethical Frameworks and Guidelines**: Robust policies needs to be implemented for responsible AI usage. These frameworks are regulation of policies, governance, education for transparency, accountability, and the mitigation of biases (Barnes & Hutson, 2024).
6. **Potential Inequities in AI Implementation**: For achieving education strategic frameworks by collaborating between policymakers and educators (Patel & Ragolane, 2024).
7. **Generating AI Applications**: Having training and awareness for responsible and ethical use (Abdullah & Mohd Zaid, 2023).

Stakeholders can improve the benefits of AI by creating an inclusive and sustainable educational environment (Al-Zahrani & Alasmari, 2024).

2.4 Policy, Governance, and Strategic Frameworks for AI in Higher Education

AI can improve engagement and learning outcomes by having learning systems according to their needs (George & Wooden, 2023; Ghimire et al., 2024). For using AI in education policy should have ethical values such as transparency, fairness and accountability in education (Drach et al., 2023). Higher education should develop new models to use AI for supporting proper decision-making (Filgueiras, 2023). This shift should have a forward-thinking governance approach which should be adaptable, inclusive and sustainable (Leal Filho et al., 2020). To safeguard academic integrity policies should be well-designed for the responsible use of AI in both teaching and research (Popenici et al., 2023; Qadhi et al., 2024). For using AI, efficient leadership is required to address the resistance and limitations of resources by reducing risk (Ghnemat et al., 2022; Tarisayi, 2024) (Table 1).

3 AI Integration Framework for Higher Education to Achieve SDG 4

AI Integration Framework for Higher Education Towards SDG 4 Achievement in Higher Education

This model consists of seven components—policy and governance development, capacity building and leadership, infrastructure and resource allocation, ethical and inclusive design, implementation programs, monitoring, evaluation and impact assessment, international collaboration and knowledge sharing (Fig. 1).

Its key impact areas are for educational quality, equitable access, personalized learning, inclusive design, lifelong learning and global collaboration as shown in Fig 1 above.

1. Policy and Governance Development

   Develop inclusive policies with ethical values, transparency, and accountability while using of AI applications. Creating governance appropriately and be inclusive, participate with stakeholders and have regulatory oversight in continuous basis.
2. Capacity Building and Leadership

   Implement continuous training for educators, administrators, and policy-makers for responsible application of AI. Inculcate leadership culture with ethical perception and strategic vision towards targeting SDG 4 accomplishments.

Table 1 Measuring the impact of AI on SDG 4 outcomes and broader sustainable development (Author)

Theme	Key findings	Limitations	Connection to (Roadmap for AI & SDG 4)	Alignment with other sources
AI applications for enhancing quality education (SDG 4.1, 4.4, 4.6)	– Personalized learning via Intelligent Tutoring Systems & adaptive platforms (Lin et al., 2023) – Improving student engagement through chatbots & Generative AI – Improved assessment accuracy & efficiency, by automating grading (Owan et al., 2023) – Resource allocation for equity (Alsagri & Sohail, 2024) – Improved teaching effectiveness by material prep, presentations, and evaluations (Nasir et al., 2024) – Lifelong learning and sustainability by proper platforms (Alsagri & Sohail, 2024)	– Checking unstructured data – Having ethical AI use – Using AI with human instructors – Need collaboration among educators, policymakers, AI developers (Akinwalere & Ivanov, 2022; Owan et al., 2023)	Directly supports the "opportunities" aspect of leveraging AI for SDG 4	Aligns well with academic on the benefits of AI in education, improving personalization, efficiency, and engagement

(continued)

Table 1 (continued)

Theme	Key findings	Limitations	Connection to (Roadmap for AI & SDG 4)	Alignment with other sources
AI for equitable access and inclusion in higher education (SDG 4.3, 4.5)	– Personalizes learning for individuals and improving engagement & achievement for diverse populations (Kohnke & Zaugg, 2025) – Helping students with disabilities via adaptive tools & intelligent tutoring (Kohnke & Zaugg, 2025) – Supporting multilingual learners with linguistic assistance & cultural representation (Davoodi, 2024) – Bridging the digital divide & providing inclusive learning environments (Edeni et al., 2024)	– Concerns about data privacy – Algorithmic bias (requires culturally responsive policies) – Institutional readiness for AI integration – Ethical considerations (transparency, fairness) crucial for mitigating disparities (Aprianto et al., 2024) – Requires institutional frameworks & policies for equitable access	Reinforces the "equity and inclusion" dimension of SDG 4 within the AI roadmap. Provides specific mechanisms (e.g., UDL principles, linguistic assistance) that AI can offer to broaden access, directly informing policies and strategies	Aligns with broader discussions on AI's to reduce educational inequalities, and emphasize need for ethical use

(continued)

Table 1 (continued)

Theme	Key findings	Limitations	Connection to (Roadmap for AI & SDG 4)	Alignment with other sources
Ethical considerations and challenges of AI in higher education	– Digital divide and accessibility: to prevent digital divide among students there should be equal distribution of AI technologies in different socio-economic groups is ensured (Aprianto et al., 2024) – Academic Integrity: Increase use of AI for automating tasks reduces the critical thinking of students and educators, these issues need to be balanced (Khatri & Karki, 2023) – Ethical Frameworks and Guidelines: Robust policies needs to be implemented for responsible AI usage. These frameworks are regulation of policies, governance, education for transparency, accountability, and the mitigation of biases (Barnes & Hutson, 2024) – Potential inequities in AI implementation: for achieving education strategic frameworks by collaborating between policymakers and educators (Patel & Ragolane, 2024) – Generating AI applications: having training and awareness for responsible and ethical use (Abdullah & Mohd Zaid, 2023)	– Requires comprehensive and ongoing efforts to create inclusive & sustainable educational environments. – Stakeholders must address concerns thoughtfully to maximize benefits and minimize risks (Al-Zahrani & Alasmari, 2024)	Crucial for the "pragmatic assessment of obstacles" and "mitigation strategies" in your roadmap. These challenges directly influence the "how" of ethical AI integration to ensure SDG 4 is achieved equitably	Highly aligns with the prevalent critical discourse surrounding AI ethics in all domains, with a strong emphasis on education-specific challenges

(continued)

Table 1 (continued)

Theme	Key findings	Limitations	Connection to (Roadmap for AI & SDG 4)	Alignment with other sources
Policy, governance, and strategic frameworks for AI in higher education	– AI can improve engagement and learning outcomes by having learning systems according to their needs (George & Wooden, 2023; Ghimire et al., 2024) – AI in education policy should have ethical values such as transparency, fairness and accountability in education (Drach et al., 2023) – Higher education should develop new models to use AI for supporting proper decision-making (Filgueiras, 2023) – This shift should have a forward-thinking governance approach which should be adaptable, inclusive and sustainable (Leal Filho et al., 2020) – To safeguard academic integrity policies should be well-designed for the responsible use of AI in both teaching and research (Popenici et al., 2023; Qadhi et al., 2024) – Using AI, efficient leadership is required to address the resistance and limitations of resources by reducing risk (Ghnemat et al., 2022; Tarisayi, 2024)	– Cultural and resource challenges in implementation – Ensuring ethical considerations are truly embedded in governance – The need for adaptation due to AI	Directly forms the backbone of the "roadmap" by outlining the necessary policy, governance, and leadership components required for successful and responsible AI integration towards SDG	Aligns with best practices in technology adoption and organizational change management, emphasizing the critical role of robust governance and ethical frameworks

Fig. 1 AI integration framework in higher education (Author)

3. Infrastructure and Resource Allocation

   Provide strong digital infrastructure and competition-friendly access routes with resources for financing AI adoption. Enter into the current infrastructural gaps to prevent any form of digital exclusion and truly co-create inclusiveness.
4. Ethical and Inclusive Design

   Make sure that AI technologies are designed with ethical values and Universal Design principles to avoid any bias and truly increase accessibility for a wide range of learner groups-those are from marginalized communities, persons with disabilities, etc.

5. Implementation and Pilot Projects

   Implement and pilot AI-based interventions aimed at personalized learning, administrative efficiency.
6. Monitoring, Evaluation, and Impact Assessment

   Establish metrics to evaluate AI's impact in educational quality for equity in education. Check the outcomes so that adjustments in policy and changes in accordance with SDG4 will be made.
7. International Collaboration and Knowledge Sharing

   Foster global partnership for sharing practices and standards, by tackling cross-border issues. In the same time collaborative research shall upgrade the said AI application in sustainable education.

4 Discussion

Application of Artificial Intelligence (AI) in higher education has opportunities and challenges in advancing Sustainable Development Goal 4 (SDG 4). AI facilitates personalized learning, by providing access for diverse students, and increasing efficiency, with inclusive and sustainable quality education using AI. The ethical concerns like algorithmic bias, data privacy, and accountability, should have an effective framework for governance. The digital divide and technological disparities is a challenge. An equitable access is required, risking the reinforcement of prior inequalities. Institutions should have inclusive policies, stakeholder engagement, and transparent ethical standards in AI applications. Systematic approach is required for building short-term policies with ethical guidelines as medium-term goals. This roadmap will help to harness full potential of AI responsibly. Responsible AI use is a priority for an innovative and equitable system contributing to SDG4 for fulfilling societal development goals.

5 Conclusion

Application of AI in Education has is significant for improving on SDG 4by fostering an inclusive, equitable and quality education. While addressing the challenges of ethical concern, digital divide and disparity in infrastructure, its essential to use AI potential responsibly with short, medium and long-term goals. To have an equitable AI use, there should be an effective governance framework, engagement with stakeholders and international collaboration.

Using AI responsibly can improve Higher Education dramatically into a more accessible, innovative and sustainable system. There is a need for coordinated effort by the policy-makers, educators and technologists. Benefit of AI can be ensured by the technological advancement supporting a resilient, inclusive environment of education with SDG 4 by 2030.

References

Abdullah, Z., & Mohd Zaid, N. (2023). Perception of generative artificial intelligence in higher education research. *Innovative Teaching and Learning Journal, 7*(2), 84–95. https://doi.org/10.11113/itlj.v7.137

Akinwalere, S. N., & Ivanov, V. (2022). Artificial intelligence in higher education: Challenges and opportunities. *Border Crossing, 12*(1), 1–15. https://doi.org/10.33182/bc.v12i1.2015

Alsagri, H. S., & Sohail, S. S. (2024). Evaluating the role of Artificial Intelligence in sustainable development goals with an emphasis on "quality education." *Discover Sustainability, 5*(1). https://doi.org/10.1007/s43621-024-00682-9

Al-Zahrani, A. M., & Alasmari, T. M. (2024). Exploring the impact of artificial intelligence on higher education: The dynamics of ethical, social, and educational implications. *Humanities and Social Sciences Communications, 11*(1). https://doi.org/10.1057/s41599-024-03432-4

Aprianto, R., Sadan, S., Fletcher, E., & Lestari, E. P. (2024). Harnessing artificial intelligence in higher education: Balancing innovation and ethical challenges. *International Transactions on Education Technology (ITEE)*. https://doi.org/10.33050/itee.v3i1.680

Barnes, E., & Hutson, J. (2024). Navigating the ethical terrain of AI in higher education: Strategies for mitigating bias and promoting fairness. *Forum for Education Studies, 2*(2), 1229. https://doi.org/10.59400/fes.v2i2.1229

Davoodi, A. (2024). EQUAL AI: A framework for enhancing equity, quality, understanding and accessibility in liberal arts through AI for multilingual learners. *Language, Technology, and Social Media, 2*(2), 178–203. https://doi.org/10.70211/ltsm.v2i2.139

Drach, I., Bazeliuk, O., Borodiyenko, O., Petroye, O., Bazeliuk, N., Reheilo, I., & Slobodianiuk, O. (2023). The use of artificial intelligence in higher education. *International Scientific Journal of Universities and Leadership, 15*, 66–82. https://doi.org/10.31874/2520-6702-2023-15-66-82

Eden, C., Adeniyi, I., & Chisom, O. (2024). Integrating AI in education: Opportunities, challenges, and ethical considerations. *Magna Scientia Advanced Research and Reviews, 10*(2), 006–013. https://doi.org/10.30574/msarr.2024.10.2.0039

Edeni, C., Adeleye, O., & Adeniyi, I. (2024). The role of AI-enhanced tools in overcoming socioeconomic barriers in education: A conceptual analysis. *World Journal of Advanced Research and Reviews, 21*(3), 944–951. https://doi.org/10.30574/wjarr.2024.21.3.0780

Filgueiras, F. (2023). Artificial intelligence and education governance. *Education, Citizenship and Social Justice, 19*(3), 349–361. https://doi.org/10.1177/17461979231160674

George, B., & Wooden, O. (2023). Managing the strategic transformation of higher education through artificial intelligence. *Administrative Sciences, 13*(9), 196. https://doi.org/10.3390/admsci13090196

Ghimire, A., Saha, S., Tiwari, A., Biswas, B., & Imran, M. A. U. (2024). Behavioral intention to adopt artificial intelligence in educational institutions: A hybrid modeling approach. *Journal of Computer Science and Technology Studies, 6*(3), 56–61. https://doi.org/10.32996/jcsts.2024.6.3.6

Ghnemat, R., Shaout, A., & Al-Sowi, A. M. (2022). Higher education transformation for artificial intelligence revolution: Transformation framework. *International Journal of Emerging Technologies in Learning (IJET), 17*(19), 224–241. https://doi.org/10.3991/ijet.v17i19.33309

Khatri, B. B., & Karki, P. D. (2023). Artificial intelligence (AI) in higher education: Growing academic integrity and ethical concerns. *Nepalese Journal of Development and Rural Studies, 20*(01), 1–7. https://doi.org/10.3126/njdrs.v20i01.64134

Kohnke, S., & Zaugg, T. (2025). Artificial intelligence: An untapped opportunity for equity and access in STEM education. *Education Sciences, 15*(1), 68. https://doi.org/10.3390/educsci15010068

Leal Filho, W., Frankenberger, F., Sen, S. K., Novo-Corti, I., Akib, N. A. M., Venkatesan, M., Salvia, A. L., Emblen-Perry, K., & Sivapalan, S. (2020). Governance and sustainable development at higher education institutions. *Environment, Development and Sustainability, 23*(4), 6002–6020. https://doi.org/10.1007/s10668-020-00859-y

Lin, Y., Luo, Q., & Qian, Y. (2023). Investigation of artificial intelligence algorithms in education. *Applied and Computational Engineering, 16*(1), 180–184. https://doi.org/10.54254/2755-2721/16/20230886

Nasir, M., Syukri, M., Adlim, A., & Hasan, M. (2024). Utilizing artificial intelligence in education to enhance teaching effectiveness. *Proceedings of International Conference on Education, 2*(1), 280–285. https://doi.org/10.32672/pice.v2i1.1367

Owan, V. J., Idika, D. O., Bassey, B. A., Etta, E. O., & Abang, K. B. (2023). Exploring the potential of artificial intelligence tools in educational measurement and assessment. *Eurasia Journal of Mathematics, Science and Technology Education, 19*(8), em2307. https://doi.org/10.29333/ejmste/13428

Patel, S., & Ragolane, M. (2024). The implementation of artificial intelligence in South African higher education institutions: Opportunities and challenges. *Technium Education and Humanities, 9*, 51–65. https://doi.org/10.47577/teh.v9i.11452

Popenici, S., Ani-Rus, A., Catalano, H., & Mestic, G. (2023). A systematic review of the artificial intelligence implications in shaping the future of higher education. *Educatia 21, 26*, 92–107. https://doi.org/10.24193/ed21.2023.26.11

Qadhi, S. M., Chaaban, Y., Alduais, A., & Khraisheh, M. (2024). Generative AI, research ethics, and higher education research: Insights from a scientometric analysis. *Information, 15*(6), 325. https://doi.org/10.3390/info15060325

Tarisayi, K. S. (2024). Strategic leadership for responsible artificial intelligence adoption in higher education. *CTE Workshop Proceedings, 11*, 4–14. https://doi.org/10.55056/cte.616